Camel

Animal
Series editor: Jonathan Burt

Already published

Albatross Graham Barwell · *Ant* Charlotte Sleigh · *Ape* John Sorenson · *Badger* Daniel Heath Justice
Bat Tessa Laird · *Bear* Robert E. Bieder · *Beaver* Rachel Poliquin · *Bedbug* Klaus Reinhardt
Bee Claire Preston · *Beetle* Adam Dodd · *Bison* Desmond Morris · *Camel* Robert Irwin
Cat Katharine M. Rogers · *Chicken* Annie Potts · *Cockroach* Marion Copeland · *Cow* Hannah Velten
Crab Cynthia Chris · *Crocodile* Dan Wylie · *Crow* Boria Sax · *Deer* John Fletcher · *Dog* Susan McHugh
Dolphin Alan Rauch · *Donkey* Jill Bough · *Duck* Victoria de Rijke · *Eagle* Janine Rogers · *Eel* Richard
Schweid · *Elephant* Dan Wylie · *Falcon* Helen Macdonald · *Flamingo* Caitlin R. Kight · *Fly* Steven
Connor · *Fox* Martin Wallen · *Frog* Charlotte Sleigh · *Giraffe* Edgar Williams · *Goat* Joy Hinson
Goldfish Anna Marie Roos · *Gorilla* Ted Gott and Kathryn Weir · *Guinea Pig* Dorothy Yamamoto
Hare Simon Carnell · *Hedgehog* Hugh Warwick · *Hippopotamus* Edgar Williams · *Horse* Elaine Walker
Human Amanda Rees and Charlotte Sleigh · *Hyena* Mikita Brottman · *Jellyfish* Peter Williams
Kangaroo John Simons · *Kingfisher* Ildiko Szabo · *Leech* Robert G. W. Kirk and Neil Pemberton
Leopard Desmond Morris · *Lion* Deirdre Jackson · *Lizard* Boria Sax · *Llama* Helen Cowie
Lobster Richard J. Kin · *Mole* Steve Gronert Ellerhoff · *Monkey* Desmond Morris · *Moose* Kevin Jackson
Mosquito Richard Jones · *Moth* Matthew Gandy · *Mouse* Georgie Carroll · *Nightingale* Bethan Roberts
Octopus Richard Schweid · *Ostrich* Edgar Williams · *Otter* Daniel Allen · *Owl* Desmond Morris
Oyster Rebecca Stott · *Parrot* Paul Carter · *Peacock* Christine E. Jackson · *Pelican* Barbara Allen
Penguin Stephen Martin · *Pig* Brett Mizelle · *Pigeon* Barbara Allen · *Polar Bear* Margery Fee
Rabbit Victoria Dickinson · *Raccoon* Daniel Heath Justice · *Rat* Jonathan Burt · *Rhinoceros* Kelly Enright
Salmon Peter Coates · *Sardine* Trevor Day · *Scorpion* Louise M. Pryke · *Seal* Victoria Dickenson
Shark Dean Crawford · *Sheep* Philip Armstrong · *Skunk* Alyce Miller · *Snail* Peter Williams
Snake Drake Stutesman · *Sparrow* Kim Todd · *Spider* Katarzyna and Sergiusz Michalski · *Squid* Martin
Wallen · *Swallow* Angela Turner · *Swan* Peter Young · *Tiger* Susie Green · *Tortoise* Peter Young
Trout James Owen · *Turtle* Louise M. Pryke · *Vulture* Thom van Dooren · *Walrus* John Miller
and Louise Miller · *Wasp* Richard Jones · *Whale* Joe Roman · *Wild Boar* Dorothy Yamamoto
Wolf Garry Marvin · *Woodpecker* Gerard Gorman · *Zebra* Christopher Plumb and Samuel Shaw

Camel

Robert Irwin

REAKTION BOOKS

Published by
REAKTION BOOKS LTD
Unit 32, Waterside
44–48 Wharf Road
London N1 7UX, UK
www.reaktionbooks.co.uk

First published 2010, reprinted 2022

Printed and bound in India by Replika Press Pvt. Ltd

British Library Cataloguing in Publication Data
Irwin, Robert, 1946
    Camel. – (Animal)
    1. Camels. 2. Camels – Social aspects.
    3. Camels in art. 4. Camels in literature.
    I. Title II. Series
    599.6'362-DC2

ISBN: 978 1 86189 649 0

Contents

Introduction

'Phoebe, do you believe that your favourite animal says a lot about you?'
'You mean behind my back?'
An exchange between Rachel Green and Phoebe Buffay in the US TV series *Friends*

This is a book about the one-humped and two-humped camel. Throughout the book I shall be referring to the first as 'dromedary' and the second as 'Bactrian'. Purists may object that strictly speaking 'dromedary' should refer only to a racing or pedigree camel. The nineteenth-century desert explorer William Gifford Palgrave tried to sort out the terminological confusion:

> The camel and the dromedary in Arabia are the same identical genus and creature, excepting that the dromedary is a high-bred camel, and the camel a low-bred dromedary, exactly the same distinction which exists between a race-horse and a hack; both are horses, but the one of blood, the other not. The dromedary is the race-horse of his species, thin, elegant (or comparatively so), fine haired, light of step, easy of pace, and much more enduring of thirst than the woolly, thick-built, heavy-footed ungainly, and jolting camel. But both and each of them have only one hump.[1]

'Dromedary' derives ultimately from the Greek *dromos*, meaning 'road', or 'course'. But in practice 'dromedary' is so widely used to refer to any kind of one-humped camel that that particular terminological battle has been lost. As for 'Bactrian',

Shadows of a caravan of camels cast on the desert sand. Dunhuang, province of Gansu, China.

this refers to the region of northern Afghanistan where two-humped camels were once plentiful. (But I do not think that there are any there today.) In order to remember which camel is which, it is helpful to think of the D of dromedary as lying on its side, producing a single hump, while the B of Bactrian on its side produces two humps.

This is my first book in which scientific matters are at issue. I am startled then to find how much contention, vagueness and sheer lack of research bedevils the scientific study of the camel. For example, how many stomachs has a camel got? The fascist camel vet Arnold Leese said three. The HMSO *Camel Corps Training Provisional* (1913) was firm that the there are four stomachs. Jibrail Jabbur, an authority of the way of life of the camel-rearing Bedouin, in refuting the idea that the camel has five stomachs, implied that it had four. The historian Edward Gibbon guessed five. Most modern authorities favour three. Among palaeontologists there is no consensus as to whether camels first appeared in the early, middle or late Eocene era, nor any awareness that variant opinions have been expressed. There are very different estimates about the length of time it takes a dromedary to copulate. Some authorities make the doubtful claim that the dromedary cannot copulate without human assistance. Wildly differing opinions have been expressed as to whether camels can show affection for humans, or whether they are intelligent. There are also quite a few guesses as to when and where camels were first domesticated. I have done my best to pick my way through all this and present what seem to me to be the best guesses.

What is it like to be a camel? How does the camel experience life? Does it take thought for the morrow? What is it like to live in a space that is to a large extent shaped and defined by its smells? How would it be to spend most of one's year utterly

'Le Dromadaire' from the Comte du Buffon's *Natural History* (1799–1800).

untroubled by thoughts of sex, but then to spend part of the winter violently obsessed with it? What kind of sex life is it in which the smell of urine plays such a large part? How is it that a father can be utterly indifferent to his calf? Why do camels like to spend time with other camels? What makes them responsive to songs and music? What is it in the camel that makes it so readily submissive to human commands? How much of its experience does the camel hold in its memory? Is the camel aware that one day it must die? If so what does it think death is? Do camels wonder what it is like to be human?[2] 'What song the Syrens sang, or what name Ulysses assumed when he hid himself among women, though puzzling questions, are not beyond all conjecture,' as Sir Thomas Browne observed. But I fear that the lived experience of the camel is beyond our ability to conjecture. It certainly is beyond mine. As Ludwig Wittgenstein observed, 'If a lion could speak, we would not understand him.'[3] Any careful contemplation of the animal world inevitably raises the unanswerable question, what is it like not to be human? But in the chapters which follow easier questions about the camel will be tackled.

1 Physiology and Psychology

These are the Ships of Arabia: their seas are the desarts.
A creature created for burthen. Six hundred weight is his
ordinary load; yet he will carry a thousand . . . Four days
together he will travel without water; for a necessity fourteen; in his often belching thrusting up a Bladder wherewith
he moisteneth his mouth and throat . . . Their pace is slow,
and intolerable hard, being withal unsure of foot, were it
never so little slippery or uneven. They are not made to
amend their paces when weary. A Beast gentle and tractable,
but in the time of his Venery: then, as if remembering his
former hard usage, he will bite his Keeper, throw him down,
and kick him: forty days continuing in that fury, and then
returning to his former meekness.
George Sandys, *A Relation of a Journey Began An. Dom. 1610*

'A camel is a horse designed by a committee' is a remark that has
been attributed to the car designer Sir Alec Issigonis (1900–
1988). As we shall see, it must have been a remarkably learned
committee, well up in anatomy, temperature control, nutrition
and desert ecology, among much else. Those committee men
would have been designers of real brilliance, for it has been estimated that fourteen per cent of the world's surface is desert and
the camel is perfectly adapted to that environment. A horse
would swiftly perish in the sort of environment in which the
camel thrives.[1]

Camels are artiodactyls (even-toed), of which there are three
types – the *Suidae*, the Ruminantia and the Tylopoda. The
Ruminantia ruminate – that is, they regurgitate the food and
chew it. In its regurgitated state it is called 'cud'. Cattle, goats,
and deer are Ruminantia as they chew cud. Although camels
and lamas also ruminate, they are not classed as Ruminantia, but
are placed in the sub-order of Tylopoda, distinguished by their

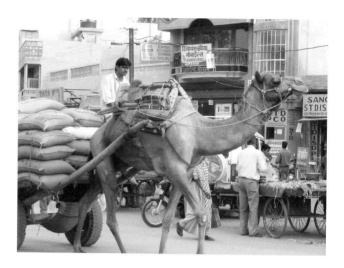

padded feet. In prehistoric times there were numerous types of Tylopod, but they have become extinct and the camels and the camelids of South America are the only surviving Tylopods.

Camel bones are dense and hard and sometimes used as a substitute for ivory. The embryos of both the dromedary and the Bactrian have the beginnings of two humps, but in the case of the dromedary these fuse into one during foetal development. (This suggests that the one-humped camel evolved from a two-humped breed.) The Bactrian hump tends to flop, whereas that of the dromedary is firmer. The average height of a dromedary is six feet (183 cm) at shoulder and seven feet (213 cm) at the hump. The Bactrian is shorter, stockier and has shorter legs. The Bactrian has a thick shaggy coat from October to March.

The female dromedary has no mane. The skull is equine shaped. There is quite a lot of disagreement about how much intelligence is contained in that skull. There is also debate about whether camels can form affectionate relations with humans.

The two-humped Bactrian camel.

Some have claimed that camels are not intelligent and, moreover, that they do not bond with men. The nineteenth-century desert explorer William Gifford Palgrave offered an unappealing account of the camel:

> He is from first to last an undomesticated and savage animal, rendered serviceable by stupidity alone, without much skill on his master's part or any co-operation on his own, save that of extreme passiveness. Neither attachment nor even habit impress him; never tame, though not wide-awake enough to be exactly wild.[2]

Palgrave's indictment fills two closely printed pages.

Again, according to Sir Garnet Wolseley's *The Soldiers' Pocket Book* (1882) the 'camel used in India is a vicious brute'.[3] Also appearing for the prosecution is Georg Gerster, author of a book on the Sahara: 'One only has to be present when a camel patrol is going off duty. No horseman ever subjected his mount to such a stream of abuse. The curses generated throughout the centuries by the camel's obstinacy, its look of sophistication and its sulky yellow-toothed mouth, from which about a hundredweight of saliva slavers every day, must be as countless as the grains of sand on the great dunes of the Erg.'[4] Someone unkindly described the camel as 'a cross between a snake and a folding bedstead.'

On the other hand, according to H. M. Barker, author of *Camels and the Outback*, 'A camel is one of the nicest creatures there is and yet he is continually persecuted by writers.'[5] Moreover, Robyn Davidson, who trekked across a large part of Australia with a string of camels, claims that the camel is quite intelligent:

> They are the most intelligent creatures I know except for dogs and I would give them an IQ rating roughly equivalent to eight-year-old children. They are affectionate, cheeky, playful, witty, yes witty, well-possessed, patient, hard-working and endlessly interesting and charming. They are also very difficult to train, being of an essentially undomestic turn of mind as well as extremely bright and perceptive. This is why they have such a bad reputation. If handled badly they can be quite dangerous and recalcitrant.[6]

The desert explorer Wilfred Thesiger gives the lie to the absurd claim that camels cannot feel affection or loyalty towards humans: 'I can remember another [camel] that was as attached to her master as a dog might have been. At intervals throughout the night she came over, moaning softly, to sniff at him where he lay, before going back to graze.'[7] According to Hassanein

Bey, the explorer of the Sahara: 'The qualities of a camel are seldom, if ever, appreciated on a slight acquaintance. The camel is as clever as a horse, if not more clever, and in some ways is more human.' A camel, which has suffered injury from a person, 'will bide his time before taking revenge with bites or kicks'. Also, 'the camel can become a very affectionate beast and very devoted to his master.'[8] John Hare, the expert on wild Bactrians, has stated that Bactrians are more intelligent than horses. He quotes one authority on the Chinese Bactrians that they pick up English faster than foreigners do.[9] Camels are sociable creatures and they do not like to travel without the company of other camels.

In general, camels are gentle creatures. But, as Sandys suggested, they are capable of nursing grudges. Pissing and shitting can be used to express dislike. The camels can also spit cud from the first stomach.

As we shall see, male rutting camels can be dangerous. Otherwise camels put up with a great deal without protest, except when

The bisected head of a camel 'plastinated' by Gunther von Hagens.

they are being loaded. Sir Richard Burton in his *Personal Narrative of a Pilgrimage to Al-Madinah and Meccah* (1893) described the potential difficulties:

> We had the usual trouble in loading them: the owners of the animals vociferating about the unconscionable weight, the owners of the goods swearing that a child could carry such a weight, while the beasts taking part with their proprietors, moaned piteously, roared, made vicious attempts to bite, and started up with an agility that threw the half-secured boxes or sacks headlong to the ground.[10]

Opinions differ about the attractive quality or not of the camel's face. Arguably the head raised high and the raised nostrils give it an arrogant appearance. The slack lips suggest stupidity. On the other hand, the soft eyes and the long eyelashes can be seen as appealing. According to Thesiger: 'To Arabs camels are beautiful, and they derive as great a pleasure from looking at a good camel as some Englishmen get from looking at a good horse. There is indeed a tremendous feeling of power, rhythm and grace about these great beasts.'[11]

The eyes, which are large and soft, project downwards so that the camel can see where it is treading. There is a third eyelid which is a defence against blowing sand and which can clear away grains of that sand. The thick bushy eyebrows and a double row of eyelashes also assist in this. The camel can see through its lids, which are only half opaque. The ears are small and lined with fur in order to filter out desert sand. The nose is similarly adapted to the desert environment. Its slit-shaped nostrils are protected by muscular flaps that can be closed at will. The camel twitches its nose to cool incoming air and to condense the moisture from its outgoing breath.

The lips are mobile, to some extent even prehensile. This reduces the need for the camel to stick out its tongue and thereby lose moisture. There are grooves from each nostril to the mouth. The cleft upper lip absorbs moisture from the nostrils. The lower lip, which is somewhat Hapsburg to start with, sags as a camel ages. The camel has a dula, or soft palate, technically known as a palatine verticulum. (Dula, or more correctly dul'a, derives from the Arabic verb *dala'a*, meaning 'to stick out one's tongue'.) The dula is often protruded from the mouth by the rutting male. Robyn Davidson describes it as 'a hideously repulsive pink, purple and green balloon, covered in slobber and smelling indescribably foul, that female camels perversely find attractive'.[12] The tongue is small and mobile. A fully grown camel has thirty-four teeth. The canine teeth of the male are longer than those of the female. The front teeth are razor sharp. The teeth of the camel grow throughout its life, so it needs tough branches or similar hard material to chew on in order to abrade them and prevent them from getting excessively long. The inner surface of the mouth is tough enough to allow the eating of thorny plants. The neck is long in order to assist it in grazing off trees or tall bushes.

Camels like to spend eight to ten hours in grazing. They tend to move around a lot while doing so and this has the effect of increasing variety in their diet, as well as doing less ecological damage to the plant life of the region. Food is gulped down and regurgitated and digested later. Camels need a lot of salt and this they usually get by eating certain sorts of bushes. Acacia is good for the camel's diet because of the plant's high water and salt content. Camels can comfortably subsist on the camel-thorn, salt bush and acacia that other animals will not touch. In that sense they have an ecological niche. (However, a diet of camel-thorn gives rise to intense halitosis.) They like

Elijah Walton, *Head of a Camel*, 1864, chalk drawing.

eating dates and they have no problem digesting the date stones. They will also eat locusts. According to Reuven Yagil, the camel, unlike true ruminants, does not have a properly developed third stomach.[13] However, its stomach is generally reckoned to have three compartments.

The first stomach compartment corresponds to the rumen of the true ruminant and it is where food is stored and where bacteria can break down the cellulose in hard grasses. So the stuff becomes cud to be chewed over more thoroughly later. The other two stomachs contain bacteria that will break down the chewed-over stuff yet more thoroughly and distribute it throughout the body. Water is not stored in the stomachs in any significant

18

quantity, despite claims that desperate men in the desert have slit open the stomachs of camels in order to drink the fluid they contain. In summer camels ruminate at midday and during the night. The stomachs' contents are recycled rapidly compared with true ruminants. Little protein intake is required. Camels can be quite eclectic in what they eat. One of Georg Gerster's camels 'developed an unbridled passion for shoe-soles, whether of rubber or of leather and regardless of the nails. I do not know which astonished me more, the apparent abundance of discarded shoe soles in the desert or the camel's acquired taste for the dilapidated handiwork of some oasis shoemaker.'[14] In Australia, a '1962 post-mortem on a half-grown camel near Oonadatta that had died suddenly revealed that its stomach contained a large plastic sheet and a length of copper wire.'[15]

A dehydrated camel can drink 27 gallons in ten minutes. Any other creature would die of overhydration if it attempted to drink so much, but uniquely the camel can store vast quantities of water in its bloodstream. It has oval-shaped, non-nucleated blood cells which resist osmotic variation without rupturing, so the cells can swell to twice their initial volume. Camel's blood is similar to that of reptiles. It also contains a large quantity of albumin (a kind of protein). Albumin helps the camel conserve water by increasing the osmotic pressure which keeps fluid in vascular spaces. The camel, like many creatures, drinks rapidly, as waterholes in the wild have tended to be dangerous places to linger at. The camel drinks to satisfy thirst, but it does not drink to store up water in advance. It can drink water with a higher salt content than seawater. It can go thirty days without water, as long as there is decent grazing, and it can go five to seven days without either food or water. In the case of very prolonged deprivation of water, the camel's mouth becomes so dry that it finds it difficult to eat.

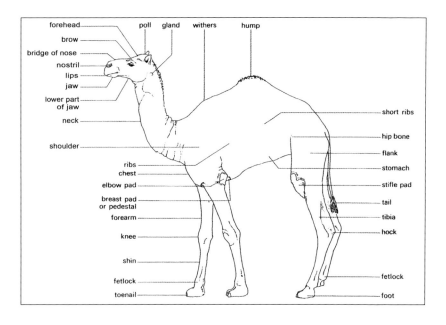

A morphological sketch of the camel.

Camel dung is rich in ammonia and therefore good as a fertilizer. The pellets are dry, neat and round. In a healthy camel the balls of shit will be as round as apples. The dry pellets are good for fires. Hence, as H.R.P. Dickson, British Political agent in Kuwait in the 1930s, observed, 'all travellers carry a bag full'.[16] The pellets shrink if the camel is dehydrated. The camel has a small bladder and it urinates in small dribbles as it walks. The urine is thick, salty and syrupy. It is also quite hot and on cold mornings Bedouin sometimes warm their hands in it. A dehydrated camel urinates less than a man. The male camel is retromingent, that is to say that it urinates backwards. (According to Arab folklore this is done out of respect for Abraham.) More plausibly, the dribble of urine cools the male camel's back legs. However, when the camel is sexually aroused, its penis reverses

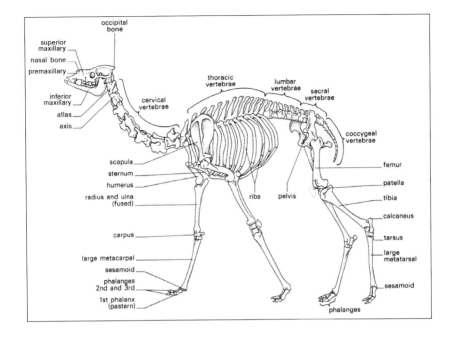

direction. The Victorian desert explorer Charles Doughty describes the camel's smell as 'muskish and a little dog-like, the hinder parts being crusted with urine; yet the camel is more beautiful in our eyes, because man sees in this creature his whole welfare.'[17]

The camel's skeleton.

The camel loses water from bodily tissues, not from the blood, and therefore there is no strain on the heart. The sweat gland is at the back of the neck. A human sweats as soon as temperature rises above the normal body temperature of 37 degrees, but the camel can raise its tolerance up by as much as six degrees before it begins to sweat. It is unique among mammals in this. In the spring camels moult and they acquire a new coat in the autumn. The coat reflects sunlight. Bactrians have two coats, a warm inner coat of down, and an outer coat which is long and hairy. Because of the hairy coat, sweat evaporates close to the skin and this helps the camel stay cool.

The humps do not store water. Fat is stored there, as an energy source. If fat were distributed throughout the body, rather than concentrated in the hump(s), it would make the camel too hot. When fatty tissue is metabolized, it produces water through reaction with oxygen. Although a fat hump is usually a sign of good health, it could just be that the camel has been eating a lot of plants that have a high water content. Apart from storing fat that can be transformed into energy, the camel's hump acts as a kind of umbrella that protects the internal organs from excessive heat. The dromedary has a single hump, rather than two, probably because that way less of the animal's surface area is exposed to heat.

A camel skeleton in Palestine, bones stripped by locusts, c. 1936.

Long legs keep the standing camel well above the hot sand. (The Bactrian, which mostly inhabits colder climes, has shorter legs.) The front legs are stronger and carry most of the camel's weight, and this is part of the reason for the swaying motion. Unlike the horse's legs, the camel's legs only connect with the body at the top of the thigh.

The camel paces – that is to say, it moves both the legs on the one side and then both on the left. This is conducive to the rolling gait. It is the only species to pace normally, though a horse can be trained to do so.

In his stylish classic of travel writing, *Eothen*, Alexander King-lake described the effect of pacing on the rider:

> The camel, like the elephant, is one of the old fashioned sort of animals that walk upon the (now mostly exploded) plan of ancient beasts that lived before the flood; she moves forward both her near legs at the same time, and then awkwardly swings round her off-shoulder and haunch, so as to repeat the manoeuvre on that side; her pace therefore is an odd disjointed and disjoining sort of movement that is rather disagreeable at first, but you soon get used to it.[18]

As noted, the camels are Tylopods, or pad-footed. But unlike most ungulates they have nails rather than hooves. The camel's nails are small. The camel does not walk upon its nails, but upon pads in front of its hooves. These leathery pads spread out when it walks. (The smaller mountain dromedary has a harder foot.) Inside the pad is a squishy sort of bladder. Because of the soft pad, the camel has a silent tread. If it walks on hard stony ground for a long time, the pads may crack and the camel will then need to wear special leather shoes. Alternatively, the

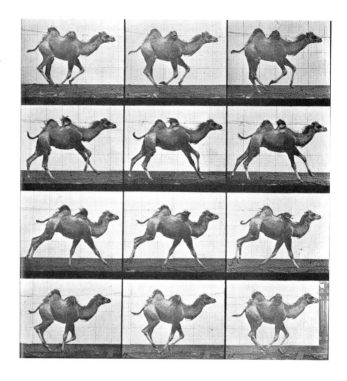

damage can be treated with tar and turpentine. When the camel
kicks, it kicks sideways. At a normal pace the camel covers three
miles an hour. At a gallop it can do twelve, or even fourteen miles
an hour. A dromedary can comfortably carry a load of six hundred pounds (273 kg), whereas the stockier Bactrian can carry
one thousand pounds (455 kg).

There is a Pushtu proverb to the effect that 'God alone knows
on what knee the camel will squat down', but it is not as bad as
all that. Still, watching a camel kneel is quite a sight. This is T.
E. Lawrence's description:

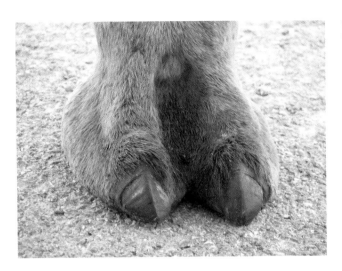

Close-up of a camel's foot.

They knelt without a noise, and I timed it in my memory: first the hesitation, as the camels looking down felt the soil with one foot for a soft place: then the muffled thud and the sudden loosening of breath as they dropped on their foreleg, since this party had come far and their camels were tired: the shuffle as the hind legs were folded in, and the rocking as they tossed from side to side thrusting outward with their knees to bury them in the cooler subsoil below the burning flints.

The camels would rest there, uneasily switching their tails till their masters remembered and looked to them.[19]

When settling down, the camel will seek out the softest spot. The brisket, also known as the chest callus, is an area of hard gristle (a keratin pad) for the camel to rest the main weight of its body on. The knees also have calluses. Camels often rest against one another to stay cool. In the daytime, they usually sit facing the sun so as to minimize the exposure of their bodies to the sun's rays. Heat from the day that has accumulated in the body is lost in the cool of the night. There are selected spots in the desert where the camels like to roll in the soft sand. These sites of dust baths are identified by sniffing. Sometimes there are communal sites where lots of camels will roll together. The main purpose of such rolling seems to be to dislodge ticks, though it perhaps also relaxes the muscles. Occasionally a female camel will roll on the ground as part of a courtship display.

There is no sexual interest in the summer and it is therefore safe to let males graze with females at that time. The male becomes sexually mature at six. His right testicle is slightly larger than the left. (Not many people know this.) It is common practice to castrate most male camels in order to prevent fights and aggressive rutting behaviour. Just a few stallions are spared for breeding. The camel's rutting is comparable to the *musth* of the Asiatic elephant. A rutting camel is known in Arabic as the *hadur*, or braying one. A sticky, smelly substance is secreted behind the ears when rutting. A male in rut is too agitated to eat much and his condition declines. Herders often give special food to the rutting male in order to make good the weight loss. He froths, gargles, grinds his teeth and is prone to violence. He urinates and uses his tail to swish his piss about. All this spitting, drooling and urinating – wastage of water – can be seen as a form of conspicuous display intended to impress the female – the male camel's equivalent of the peacock's tail. There are occasional fights between rutting males for ascendancy. As noted, the blowing out of the soft palate (dula) seems to impress the female no end. Its smell is attractive to her. (The Bactrian, however, does not have this mouth flap.) For his part, the male sniffs at female genitalia and he is roused by the smell of the urine of a non-pregnant camel.

The female dromedary needs ovulation to be stimulated by mating. In the northern hemisphere copulation mainly takes place between November and February. The female wags her tail at the approach of the male and displays her genitals. Often, when she is ready for sex, she sits down spreading her hind legs and urinates. But if she is reluctant, the male may put pressure on her neck to force her to sit down. Bertram Thomas, the explorer of Arabia's Empty Quarter, has a footnote on copulating camels:

> Camels are thus, like the llama and the lion, rare in the animal kingdom in the performance of the act in the sitting position. The Badu master is necessary to the operation, scooping the sands round the cow's legs for her comfort, inserting the penis – the formation of which is in reverse axis to nearly all the rest of mammal creation, and interfering after a few minutes to drive the bull off. After ten days if no result is apparent, the cow's master

Camels copulating, from a picture postcard.

The two stages of camel copulation.

will find another bull to serve her. The sign of pregnancy is the flag-wagging of her ridiculous tail when approached by a rider to mount.[20]

Average copulation time is five and a half minutes (but one authority says around twenty minutes!).[21] His penis is hook-shaped. Sometimes men assist the copulation – for example by tying the female's foreleg to her shoulder. But obviously human assistance is not absolutely necessary, otherwise there would not be wild camels and feral camels. A male can serve fifty or more females.

Cross-breeding between camelids is possible and in Persian a hybrid from a Bactrian and a dromedary is called a *bokt*. It has a single hump that is longer than the dromedary's hump. The *bokt* is big and strong and a good pack animal. A *cama* is the product of the mating of a camel and a lama.[22]

The female first calves at the age of five. It is very rare for twins to be born. Calf mortality is high: between twenty and forty per cent die in the first year. When about to calve the female has a tendency to wander away. Charles Doughty described the parturition in his peculiarly elaborate and archaic prose:

The yeaning camel-cow, lying upon her side, is delivered without voice, the fallen calf is big as a grown man; the herdsman stretches out its legs, with all his might: and draws the calf, as dead before the dam. She smells to her young, rises and stands upon her feet to lick it over. With a great clap of the man's palm upon that horny sole, *zôra* (which, like a pillar, Nature has set under the camel's breast to bear up the huge neck), the calf revives: at three hours end, yet feeble and tottering, and after many falls, it is able to stand reaching up the long neck and feeling for the mother's teat. The next morrow this new born camel will follow to the field with the dam.[23]

Female camels are often sterile and they often abort, sometimes owing to parasitic infections, so it can be difficult to keep up the numbers in a herd. The gestation period of a dromedary is a

A Bactrian camel suckling her calf.

little over twelve months and that of a Bactrian can be a month or more longer. Calves are suckled for three or four months. The she-camel will mother a calf for a whole year. At the end of a year the calf is weaned. The milk is low in fat and high in vitamin c, but it does not store well.

The camel has four teats. (The Bedouin customarily tie up two of them so that they can keep some milk for themselves.) The sweetest milk comes from camels that have most recently given birth, but camel milk has no cream. It is available for eleven or more months a year – unlike sheep, goats or cows, which only lactate for five months. The female mourns when her calves are taken away and will weep and mourn for a dead calf for about ten days. A *baww* is the term for a stuffed calf assembled by Arab herdsmen to ease the grief of a mother camel.

A calf can usually stagger upright in half an hour. It remains close to its mother for the first five years. Females reach puberty in the second year and start bearing calves in the sixth year. Up to the age of three a camel needs some sleep. Thereafter some claim that it does not sleep, but there is some dispute about this and the staff of Whipsnade Zoo are emphatic that camels do sleep (heads down, eyes closed, etc.). In the ninth year, with the appearance of canines, the camel's growth is complete. It has a life expectancy of about forty years.

It is possible that in navigating male camels follow the stars. The lead camel female urinates every six kilometres or so to mark the way for those that follow.

Almost all the world's camels are either domesticated or feral (feral camels being the descendants of domesticated camels that have escaped or been turned loose). Exceptionally, there are some hundreds of wild Bactrian camels in Mongolia and China. It seems that the wild Bactrian may belong to a different species from the domesticated Bactrian.

The wild Bactrian is greyer and slimmer and has smaller, more widely spaced conical humps. It has a thinner head, shorter hair and a different DNA, in that it has three more genes than the domesticated Bactrian. Also it has no chest callus and, unlike the domesticated Bactrian, the wild Bactrian can drink saltwater slush. It is conceivable that the wild Bactrian is the ancestor of the domesticated camel world-wide.[24]

Camels have a taste for music. In 1911, in an article entitled 'Les Animaux – sont ils mélomanes?', the magazine *Nos Loisirs* reported on an experiment carried out in the Bronx Zoo, New York, when a naturalist took a gramophone round the zoo and played music to various animals. Although quite a few of the animals were indifferent or even hostile, the llama stood rigid to attention to hear the music. The camel was equally delighted and tried to get its muzzle into the antique gramophone's horn and rubbed its face against it. But for many centuries before the invention of the gramophone, Bedouin have sung to camels to urge them on or to get them to drink.[25] A singing *hadi*, or

A camel listening to an old-fashioned horn record player, from the French magazine *Nos Loisirs*.

camel driver, can get camels to move faster if he sings well. According to the fourteenth-century historian Ibn Khaldun, there were teachers in Egypt who specialized in teaching *al-hida*, the camel driver's chant.[26] In medieval Persian poetry lovers were often compared to camels which bear their burdens patiently and hurry to their master's voice or flute. Camels identify their master by his songs. In the sixteenth century, Leo Africanus, in his account of the camel, mentioned the songs that are used to encourage the camel. He also described a dancing camel in Cairo. Rosita Forbes, an adventuress who travelled in the Libyan Sahara in the 1920s, remarked that 'it is a curious fact that camels walk more quickly and straighter to the sound of singing. Therefore the blacks and She-ibs drivers used to chant wild melodies of love and prowess till even my great blond beast forgot his amorous gurglings and kept his nose in a bee-line for the horizon.'[27] Mongol herdsmen make use of a mouth harp in order to get a female camel first to weep and then to accept to suckle a calf which is not hers. A variant of this procedure was filmed for *The Story of the Weeping Camel* (on which see chapter 6). Hassanein Bey gave several examples of the song that Saharan nomads sing to their camels, of which this is one:

> The sand-dunes hide many wells
> That brim with waters unfailing.
> You come to their margins like bracelets
> Wrought of gold and rare gems in far countries.[28]

Camelids have a special immune system which gives them more resistance to certain diseases than other ruminants or, for that matter, humans. However, they are peculiarly vulnerable to camel pox and trypanosomiasis. Camel pox is a viral skin disease. Lesions usually appear on the head. Young camels are most vulnerable to it and most likely to die from it. More normally though, the disease runs its course in a matter of weeks. In trypanosomiasis, also known as surra, flying and biting insects, especially tsetse flies, are responsible for infecting the blood with a parasite. This gives rise to fever, anaemia and emaciation. If untreated, it results in ninety per cent mortality. The symptoms include recurrent fever, progressive anaemia and general physical decline. Initially at least, the disease is tricky to diagnose. It is a summer disease and horses and mules are also vulnerable to it. Mange, an inflammation of the skin (in Arabic, *jarrab*), is caused by gastrointestinal worms and is potentially lethal. It can kill a camel in a matter of weeks. It usually arises out of lack of fresh grass and it steadily weakens a camel. After trypanosomiasis, it is the biggest single camel-killer.[29] T. E. Lawrence describes how the tribesmen he was with tried to treat it by rubbing the camel's coat with butter.

Some camels get afflicted with the mysterious bent-neck syndrome in which the camel is hardly able to lift his head above the ground. Rabies is fairly rare, but it does occur and it is lethal. Tuberculosis is rare, but camels can catch it from cattle. Sedentary camels are particularly prone to ticks. Lawrence described how humans too could be plagued by these creatures:

The camel-ticks which with blood from our tethered camels had drunk themselves into tight slaty-blue cushions, as thick as a bean, and as large as a thumbnail, used to creep under us for warmth, hugging the animal-like leather underside of the sheepskins: and if we rolled on them in the night (for men sleeping on the flat lay at first usually for softness on their faces) our weight would burst them into brown mats of dry blood and dust.[30]

A camel tick.

Camel ticks tend to cluster under the tail. Ticks should be removed daily. If you pick out the ticks and throw them on the fire, they will explode like popcorn. Such fun! (But the camels themselves do not enjoy having the bloodthirsty ticks pulled off.) Other skin diseases arise from lack of salt. No research has so far been undertaken into mental health of camels, though John Hare has noted the occasional tendency of male wild Bactrians to go mad and gallop for days and nights until they drop dead.[31]

Apart from the proneness of camels to camel pox, trypano-somiasis and, perhaps, insanity, the camels seem extraordinarily well designed in every aspect of their anatomy and physiology for surviving in a desert environment. So much is this the case that creationists like to adduce the camel as a perfect example of God's creative handiwork and as something that could never have come into being over millions of years of evolutionary selection. As Mark Stewart, writing for the East Tennessee Science Creation Association, put it: 'The camel and its specialized water

conservation features demonstrate incredible design. It is difficult to imagine how all the various features the camel needs in
order to survive could have developed by gradual evolutionary
processes.'[32] However, the argument from lack of imagination is
not a strong one.

2 Ancestors of the Camel

The hot dusty Jurassic period and the warm and humid Cretaceous period, about 206 to 65 million years ago, were the heyday of the giant dinosaurs – and of conifers, ferns and cyads (palmlike plants), though by the end of the Cretaceous period flowering plants had replaced the conifers, ferns and cyads in many regions. Then, some 65 million years ago, for reasons that are hotly debated, there was a massive range of extinctions in which about half of the species that had flourished till then became extinct, including all the dinosaurs, as well as many sea creatures and reptiles. Only smaller creatures survived. The mass extinction was to open up all sorts of ecological niches for mammals.[1] In the Palaeocene era which followed much of the former tropical flora disappeared. But new plants emerged and in the Palaeocene period even the polar regions were covered with forests.

At the very end of the Palaeocene period new types of mammal appeared for the first time, including hoofed mammals, rodents and primates. Around 55 million years ago the Palaeocene era gave way to the Eocene, which was to last until 34 million years ago. ('Eocene' is derived from the Greek *eos*, meaning dawn, and *kainos*, meaning new.) The climate then in North America was warm – what there was of it, that is, for much of present North America, especially in the east, was actually under water. In the early part of this epoch, crocodiles had basked in the Arctic Circle. In this era mountain ranges running north–south in western North America were formed. Grasses first appeared in the Eocene, then the area covered by grass expanded vastly in the Oligocene period, 34 to 24 million years ago.

As temperatures cooled, true savannah (grassland with scattered tree clumps) extended. This increase in large grassy areas allowed grazing mammals and their predators to grow in size. Tapirs, rhinoceroses and tiny horses were among the mammals that appeared. Large horned beasts like the Uintatherium and the Tinoceras roamed western America, though most such creatures were to die out by the end of the Eocene epoch. The division of mammals into odd-toed and even-toed also occurred in this period.

THE COMING OF THE CAMEL

There are no camels today in North America, except for those in zoos or on the farms of eccentric ranchers, but camelids spent the first 36 million years of their existence in America. In its earliest form the camel probably appeared in the late Eocene epoch, that is to say, about 40 million years ago. At this time the climate was cooling and the tropical forests had receded. There is, however, disagreement on this point and some palaeontologists claim that the earliest camels appeared in the middle or even the early Eocene.[2] At the very end of the Eocene period Antarctica started to freeze and the Earth's temperature started to cool. A large number of species perished as a result.

The earliest of the camelids, *protylopus*, was no larger than a hare, its front legs were shorter than its hind legs and it had

A reconstruction by Michael Long of *Poëbrotherium*, a very early (Oligocene) camelid, found in North America around 30–38 million years ago.

low-crowned teeth. It probably grazed on low-level forest vegetation. As the name may suggest, it was a Tylopod – that is to say, it was even-toed, as it had four toes. Later the camelids lost their side-toes, some time during the Oligocene and the Miocene epoch (24 to 5 million years ago). The earliest types of the camelids probably had no humps and probably resembled llamas. In the Oligocene epoch some camels developed the long necks that allowed them to browse off trees and the tops of shrubs. In America long-necked camels occupied the same niche as did the giraffe in Africa.

The continuing cooling of the earth led to the widespread replacement of savannah by steppe lands and slowly, during the Miocene epoch and the Pliocene epoch (5 million to 1,800,000 years ago) camelids grew larger, their necks and legs grew longer, their distinctive dentition evolved and so did the soft foot that is so characteristic of the modern camel. As to the dentition, hypsodonty (lengthening of teeth) increased among camels,

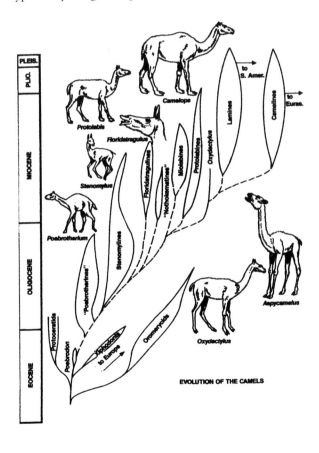

The evolutionary tree of the camel.

A reconstruction by Michael Long of *Aepycamelus* (aka *Alticamelus*), a giraffe-like camelid found in North America around 5–21 million years ago.

horses, rhinoceroses and other creatures. These teeth with higher crowns were desirable for chewing on scrub and the accompanying grit. (Modern camels do not have such high-crowned teeth.) Camels also developed a pacing gait. All sorts of camelids appeared that were not destined to survive until modern times, such as the *Oxydactylus* (a camel with a neck like a giraffe), *Stenomylus* (a miniature gazelle-like camel) and *Alticamelus*, also known as *Aepycamelus,* which at nineteen feet (5.8 m) tall was the earliest of the giant camels. It fed off the tree canopy and somewhat resembled a giraffe.[3]

PLIOCENE AND PLEISTOCENE CAMELS

In the Pliocene epoch temperatures became more varied. North America and South America, previously separate continents, collided 2.5 million years ago, creating a land bridge that allowed the camelids to migrate southwards. At almost the same time a land bridge formed which allowed animals from America to cross the Bering Straits into Asia. Over the millennia divergences developed between those camelids that migrated to Asia, the ancestors of the Bactrian and the dromedary, and those that moved down into South America, the ancestors of the vicuña, guanaco, lama and alpaca.

The Ice Age began at the start of the Pleistocene epoch, some 1,800,000 years ago. More species of mammals flourished in the Pleistocene period than exist today. *Camelus,* the sort of camel we know today, first appeared about half a million years ago in North America. Probably it was two-humped. But there were other sorts of camels. It seems probable that the one-humped camel evolved from the two-humped one, though there is no evidence of this in fossil remains. However, the foetus of a one-humped camel goes through a two-humped stage.

The camel is a timid creature and is poorly equipped for defence from the predators. Its best defence has been to retreat to desert or mountainous areas in which the predators would have found it difficult to survive. The dromedary and the Bactrian found their place in the desert. In South America, the camelids, comprising the llama, vicuña, alpaca and guanaco, specialized in grazing at high altitudes.

A reconstruction
by Michael Long
of *Titanopylus*,
a camelid found
in North America
around 300,000
to 1 million
years ago.

In the first part of the Pleistocene epoch large camels flourished on the savannah. The giant *Titanotylopus*, whose remains have been found in Nebraska, appeared in the Pleistocene ice age, as did *Megatylopus*. There was a general trend towards gigantism in this era and there were at least eight types of giant camel. Large camelids flourished as far north as Alaska. Very large camel bones have been found in the Yukon that resemble those of *Titanotylopus*. The Alaskan camel had a narrow skull with not much space for a brain. It was about 3.5 metres tall and had a hump. The remains of a slightly smaller camel, the *Camelops* (literally 'camel face'), a creature between camel and llama, has also been found in Alaskan and Yukon Ice Age deposits. About 3 million years ago the last Ice Age began and continued into historical times. It is possible that camels survived in Alaska until the peak of this last glaciation, about 25,000 years ago. Late Pleistocene glaciation meant hard times for camels. Most of America's megafauna (large animals) vanished in the late Pleistocene.[4]

It is odd that the *Camelops* disappeared when, after the Pleistocene era, the earth was getting warmer and arid areas suitable for camels were increasing. But there was a huge number of extinctions in the late Pleistocene, especially of large land mammals. It is possible that men hunted them to extinction. The western camel (*Camelops hesternus*, literally 'yesterday's camel') was certainly hunted in America. Camel bones with slaughtering marks on them have been found at two archaeological sites. The western camel was a little taller than the modern Bactrian, but there is no telling how many humps, if any, it had. It became extinct about 10,000 years ago. Horses similarly became extinct in America, until they were reintroduced by Europeans in the sixteenth century.

The first camels to enter Asia seem to have been Bactrians. The one-humped dromedary probably evolved later, very probably in the Arabian Peninsula. Bactrian-like camels stayed in Asia, as well as migrating westwards as far as southern Russia. The remains of a *Titanotylopus*, or *Giganotylopus*, have been found in southern Ukraine, which it plausibly reached about 5 million years ago. Ancestors of the dromedary migrated to the Middle East and North Africa. There were camels in prehistoric Africa (and they feature in cave frescoes), but having become extinct there, they were reintroduced, probably in Roman times. In prehistoric times the wild dromedary in the Middle East became extinct and the species survived only as a domesticated animal.

Camelus moreli and the modern camel (green silhouette).

Very few fossilized camel remains have been found in the Old World except for a few Pleistocene fragments. But recently, in 2005–6, the fossilized remains of a giant camel were found in the Palmyra desert in central Syria.[5] The dromedary, which was twice the size of the present-day camel, as tall as a giraffe or elephant, died about 100,000 years ago, and seems to have been killed while drinking at spring. Since human remains have been found in the same area, the camel could have been killed by men, though it is not clear whether those men would have been of the *Homo sapiens* or Neanderthal species. Hitherto it was not thought that there were dromedaries in the Middle East more than 10,000 years ago. Additional fragments of bone in other layers of rock suggest that camels had been in the region for thousands of years. Although the region is a desert today, it would have been savannah grassland 100,000 years ago. Modern camels are thought to have entered the Middle East some 6–7,000 years ago.

Because camels roam in barren, inhospitable areas, they have acquired little experience of fighting off predators. Like dolphins or certain types of reindeer, camels are what expert on

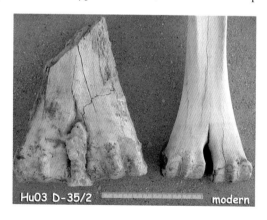

The foot of a now-extinct giant camel, *Camelus moreli*, living in Syria around 100,000 BC, discovered in 2005–6; and the foot of a modern camel.

A reconstruction of *Camelus moreli*. The modern-day camel stands 7 feet (2.1 m) tall at the top of its hump; *Camelus moreli* was around 9 feet (2.7 m) tall.

the history of camels Richard Bulliet describes as 'tame in the wild' and therefore easy to domesticate. Stephen Budiansky argued in his 1992 *Covenant of the Wild: Why Animals Chose Domestication* that during the Ice Age, camels, like dogs and horses, ensured their survival by becoming domesticated. Protected and fed by humans, camels had a better chance of surviving than if they foraged independently in the wild. Budiansky argues that 'in an evolutionary sense domesticated animals chose us as much as we chose them.' Neoteny, the prolonged retention of immature characteristics or features in the adult form, tends to be an aspect of domesticated animals. Thus they retain the curiosity, trustingness and desire for affection that is characteristic of puppies, kittens or calves.

3 Practical Camel

If you are thinking of keeping a camel as a pet, remember that they live up to thirty or even fifty years.[1] Camel calves are cute and playful, but do not buy one younger than six months as before then its mouth is not fully formed. A camel less than six years old cannot manage heavy loads. A fully grown camel, aged six and above, will be easier to manage. If you are buying a camel for riding, a female is preferable. The bulls can be unpredictable and violent, unless they are castrated. A pedigree she-camel should have sharp-pointed ears, strong shoulders, an upright neck, prominent shoulder blades, a broad chest and small feet. According to the Army manual *Camel Corps Training Provisional 1913*, 'A high-domed forehead is a sign of good breeding.' Avoid camels with high withers.[2] ('Withers' refers to the ridge between the shoulder blades.)

Other sources provide additional pointers. The softer the fur, the higher the quality. A large hump makes fitting the saddle difficult. A bloated hump suggests that the animal has been watered too often and fed too much green food. Check the nostrils for scar marks, as this may mean that several nose reins have been previously broken by a fractious animal. Check the coat, looking particularly for burn marks, which may mean the camel has been subjected to the traditional remedy for arthritis or bad sprains. Avoid camels that have been cauterized or have bald patches (as both are signs of treatment for previous sickness). Watch out for ingrowing toenails. Make sure the animal is not blind; the eyes should be bright. Check the droppings, which should be slightly moist and dark green, and whether the animal urinates easily. It helps to look confident when inspecting the beasts. Justin Marozzi (who travelled with camels and a friend across the Libyan Sahara) gives tips on how this is done:

> Together, we walked slowly around each one, trying to look as though it was the most normal thing in the world. We bought camels all the time of course. We nodded sagely, conferred and shook our heads regretfully (we did not want to look too eager), checked their legs (all of which to our untrained eyes looked crooked), stared into

Camels in a paddock, Dubai.

their eyes, slapped their flanks with what we hoped was
the air of connoisseurs and pondered.[3]

If you are keeping a camel in the UK, you will need a Dangerous
Wild Animal License.

RIDING

Now how to ride a camel. Before saddling up, use a rake to comb
the top of its coat to get rid of stones or twigs. 'Hoosh' seems to
be the most commonly used word for getting a camel to kneel.
Geoffrey Moorhouse, who in *The Fearful Void* (1974) gave a vivid
and harrowing account of his ultimately unsuccessful attempt
to cross the Sahara from Mauritania to the Nile, provided the
following advice on mounting a sitting camel: 'place your right
foot on his back in front of the saddle, but behind the beginning
of the neck, to spring up, and make a half turn to the left in the
process, landing neatly in position with both legs astride the
pommel'.[4] Easy. In addition, I would advise you, once in the

saddle, to lean back and be prepared for the camel's rear to rise first, which it does quite abruptly. Then lean forward as the rest of it rises. (Again, when the camel goes down, you should first lean backwards, as the camel first goes down part of the way on its front legs, then all of the way on its back legs and then the rest of the way on its front legs.) As it moves off, the camel, unlike the horse, paces (that is to say that the legs on the same side are lifted together and hence it sways a lot). You will be rocked forwards and backwards. The normal speed of a camel is three miles an hour. Do not slouch or you will get backache. Use your feet more than the head-rope (which is attached to a peg in the nose) in order to direct your camel. Use a riding stick on the camel's rump if you want it to speed up. Walking a camel is harder on your back than travelling at a trot. Once you are relaxed and used to the rhythm of the camel's pacing, you can read while riding if you want to. One can cover thirty or forty miles a day fairly comfortably. Since the camel's bladder is small, frequent halts are necessary (say, every half hour). Do not over-water your

The author on a camel in Oman.

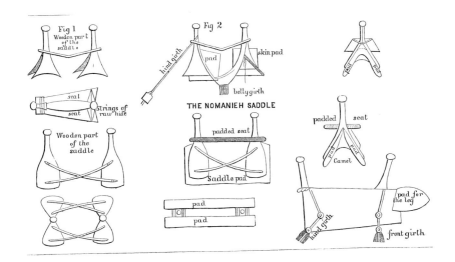

Fig 1
Wooden part of the saddle
seat
seat
Strings of raw hide
Wooden part of the saddle

Fig 2
hind girth
pad
skin pad
belly girth

THE NOMANIEH SADDLE

padded seat
Saddle pad

pad
pad

padded seat
pad
pad
Camel

pad for the leg
hind girth
front girth

North Arabian camel saddle from an 1857 US *Report of the Secretary of State for War . . . respecting the Purchase of Camels for . . . military transportation.*

camel; every third day will do. Camels like to drink when the sun is properly up, but it is best to let the camels graze early in the morning, before the temperature gets so hot that the camels become more concerned to find shade than food. The pack saddle has to be regularly adjusted as the size of the hump changes.

Sir Garnet Wolseley's *Soldiers' Pocket Book* (1882) warns of the dangers of riding your camel over rough ground or overworking it:

> After rain, in clay soil or over rocks or stony places, they split up and they are consequently useless there . . . They are extremely delicate in constitution and liable to diseases little understood. When suffering from overwork they do not recover with rest like horse or mule: they pine and die away. They require a long time to feed, at least six hours; owing to their great height they suffer severely from ill-balanced loads.[5]

50

If you are putting a load on a camel's back, you need patience and concentration to set the load evenly so that it will not unbalance the camel or chafe it. Nomad camel men commonly take up to an hour getting the load right. As Chingiz Aitamov has one of his characters remark in his novel *The Day Lasts More Than a Hundred Years*, 'To saddle a camel is no small task; it is rather like building a house.'[6]

RESTING

Do not overwork your camel. As Gerster pointed out, 'the camel has its own method of protesting against unreasonable treatment: it dies.'[7] When your ride is finished, tap the camel on the neck to make it kneel. It is likely that after the first three days of riding you will have a very sore bottom. Assuming that you are

Camels being watered in Palestine in the 1930s.

somewhere hot in Asia or Africa, travel early in the morning and then again early in the evening. Frederick Burnaby, cavalry officer and author of *A Ride to Khiva* (1876), advised that camels will only feed in daytime and so it is best to travel with them during the night. It is a bad idea to let camels rest in snow, as they then 'get cold in the stomach, an illness which generally proves fatal to them'. According to Burnaby there 'is a good deal of nonsense talked and written about the patience and long-suffering of these so-called ships of the desert'. He found them prone to run away or suddenly lie down without any forewarning.[8] Halt in daylight to choose good ground for barracking: that is, with no sharp stones, if possible. (To barrack is to get a camel to sink to its knees.) Be careful tethering the camel, as it can use its front lip to untie knots.

DIGESTIVE MATTERS

Give the camels time to graze before sunset. Camels like variety in their diet. The thornier the feed the better, though a camel's thorn

Camels drinking in Palestine in the 1930s.

diet results in extreme halitosis. If the camel is constipated, give it Epsom salts; beans might also work, as they tend to give camels diarrhoea. The camel needs lots of grazing time, ideally eight hours per day, rather than Wolseley's recommended six hours. The lead camel urinates every six kilometres or so so that the scent can guide the following camels. You can use camel urine to wash your hair. It will protect your hair from nits and give it a reddish tinge. If you and your camel are lost in the desert and you have run out of water, you can turn to drinking camel urine. You can also cut out the camel's cud in order to squeeze liquid out of it.

NOSE PEGS

One cannot use a bit on a camel, because it chews all of the time. Therefore the insertion of the nose peg is essential for guiding a camel, but the insertion is a gruesome business. Robyn Davidson describes two methods of doing it:

A camel nose-peg and punch.

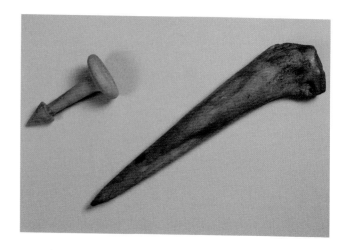

Sallay skewered the flesh straight through from the inside with a sharpened mulga stick, then inserted the wooden peg into the whole and dressed it with kerosene and oil. Kurt's method was more sophisticated if not better. He would mark the spot on the nose with a marker pen, punch a small hole in the flesh with a leather punch, widen that hole with a butcher's skewer driven through from the inside up to the hilt, and follow this with the insertion of the peg, which by the way, looks more than anything else like a small wooden penis.[9]

If there are scars left by old nose rings this suggests that the animal is difficult to control.

BE CAREFUL

Watch out for a camel's kick. It can kick sideways, forwards and backwards. If attacked by a rutting camel, you should rip off

A Bactrian camel displaying the dula, an inflatable sac in the mouth which can be used to make a noise during rut.

your clothes and throw them before him. He may accept this as propriation. As David Taylor, a zoo vet, observed: 'In Asia when a camel driver senses trouble, he gives his coat to the animal. Rather like Japanese workers reported to work off frustrations by beating up models of their executives, the camel gives the garment hell – jumping on it, biting it, tearing it to pieces. When the camel feels it has blown its top enough man and animal can live in harmony again'. Bimbashi MacPherson, who was commissioned into the Egyptian Camel Transport Corps during the First World War, offers the following advice if you are charged by an angry camel, 'calmly await the on-slaught, step aside and smite the brute on the nose with the butt end of a whip – that would stop and daze them – then a blow over the knees would bring it down, so that its legs might easily be tied'.[10]

Camel dung burns well because of its high cellulose content. It is also high in ammonia and therefore good as fertilizer. In *The Seven Pillars of Wisdom*, T. E. Lawrence wrote evocatively about camel shit in the desert:

> There were no footmarks on the ground, for each wind that blew swept like a great brush over the sand-surface stippling the prints of the last travellers till the surface was again a virgin pattern of innumerable tiny waves. Only the dried camel droppings, which were lighter than the sand and round like walnuts, escaped its covering. They rolled about between the ripples, and were heaped up in every corner-pocket by the driving wind. It was perhaps by them, as much as by his unrivalled road-sense, that Auda knew his way.[11]

There is an Arab saying: 'The track cannot lie'. *The Arabian Nights* 'Story of the Sultan of al-Yaman and His Three Sons' is a tale about the young men's abilities as detectives and in particular their skill at deducing an enormous amount from the tracks of a camel. This story is also found in the histories of al-Tabari and al-Mas'udi. It is classified by folklorists as tale-type 'AT655: The Strayed Camel and the Clever Deductions'.

As Michael Asher notes: 'To live and survive in the desert, one has to learn the desert's grammar – the syntax of the wind, stones, plants and animals.'[12] The speed of a camel can be determined from the spacing of its tracks, as the distance between the fore-print and the hind-print will tell a tracker how fast the animal was travelling. André von Dumreicher, who saw service with the Egyptian Coastguard Camel Corps in the early twentieth

'Courier of the Desert', an engraving based on painting by Horace Vernet, in the 1857 US report on the military use of camels.

century and who frequently tracked down hashish smugglers, provides useful detail here: a tracker

> will be able to estimate with great accuracy the pace of the camels he is chasing, by noticing the juxtaposition of the footprints. Thus, if a full-grown camel ambles at the rate of five miles an hour, the claws of the off-hind foot

reach up to the level of the back of the near-fore; if the pace is increased to seven or eight miles, both near-fore and off-hind will draw up to the same level, and if the pace be increased it will be safe to add one mile more an hour for every two inches, by which the off-hind overreaches the level of the near-fore. For instance, if the distance between the near-fore and off-hind is twelve inches, one can conclude that the camel is going at the very great speed of about fourteen miles an hour. A camel going at this speed will throw back, at the same time, a considerable amount of sand.[13]

In Von Dumreicher's experience (which he compared to that of 'a boy's adventure story') an innocent caravan, transporting dates say, will allow the camels to meander and graze, whereas hashish smugglers will drive their camels on to follow a straight route. The straighter the trail, the more suspicious it is.

A camel that has been let free to wander like to 'drink the breeze', and only if the wind (and accompanying sand) reaches a certain strength will it turn its back and move in the same direction as the wind.[14] Tracks of camels by night are less straight and the camels are more likely to go over stony ground. The appearance of a male camel's footprint will follow that of its father. With time, the outlines of a camel's hoof-print becomes less clear. In early morning, when dew is falling, more sand is thrown out by the hoof print. For two days the track is reddish in appearance. Camel droppings generally retain their moisture for the first day. If they are fresh, one can smell the herbs in the dung. If the camel is in good condition, its droppings will be symmetrically round. If the droppings are green they suggest that the camel has been eating clover. If the camel is thirsty, its droppings are harder, smaller and date-shaped. In the case of

A camel foot-print.

fresh droppings, one can smell if the camels have been well fed on odoriferous plants.

A camel-rearing Bedu read a camel's track as easily as most people read a face and, for example, can deduce the identity of a mature camel from the memory of that camel's print when it was two years old. The Al Murra of Saudi Arabia have long had the reputation of being the best trackers. Acording to Dickson, a Political Agent based in Kuwait, 'it was easy for a Murri to say whether the tracks of a single camel were those of a white camel (*wadha*), or a brown camel (*hamra*), or a black camel (*sanda*) and still easier to say it was male or female, in calf or not'.[15] The anthropologist Donald Powell confirms the Al Murra's skill in tracking. All members of a *bayt* (household) know the individual footprints of their own camels and some at least of their kinsmen's camels. Al Murra trackers used to be attached to all the main Saudi police stations. *'Arifin al-kaf* are men expert in reading tracks. A *ka'if* is someone who can establish kinship

between camels primarily by the feet. They read the criss-cross grain of the camel's hoof and the droppings.[16]

The explorer Wilfred Thesiger tells of how he was travelling with some Bedouin in the Empty Quarter. They passed some scuffed tracks. An old man leapt down to examine them and then a little further on to examine some camel-droppings which he broke between his fingers. He reported of the tracks: 'They were Awamir. There are six of them. They have raided the Junuba on the southern coast and taken three of their camels. They have come here from Sahma and watered at Mughsin. They passed here ten days ago.' (All this was later confirmed in every detail.)

Thesiger continues:

Here every man knew the individual tracks of his own camels, and some of them could remember the tracks of nearly every camel they had seen. They could tell at a glance from the depth of he footprint whether a camel was ridden or free, and whether it was in calf. By studying strange tracks they could tell the area from which the camel came. Camels from the Sands for instance, have soft soles to their feet, marked with tattered strips of loose skin, whereas if they came from the gravel plains their feet are polished smooth. Bedu could tell the tribe to which a camel belonged, for the different tribes have different breeds of camel, all of which can be distinguished by their tracks. From looking at their droppings they could often deduce where a camel had been grazing and they

Brand marks on Egyptian camels, from the 1857 US report on the military use of camels.

could certainly tell when it had last been watered, and from their knowledge of the country they could probably tell where.[17]

But some caution is necessary, as tricks may be used against trackers. One of the greatest trackers in the employ of the Egyptian Frontier Police in the early twentieth century could fake individual camel tracks with his hands.[18] The Arabs fighting alongside T. E. Lawrence in the First World War loaded up with extra camel droppings and scattered them en route in order to make the Turks think that his force was bigger than it actually was.

EATING CAMEL

Since the camel does not have a cloven hoof, its meat is forbidden to Jews: 'Nevertheless these ye shall not eat of them that chew the cud, or of them that divide the cloven hoof; as the camel, the hare, and the coney: for they chew the cud, but divide not the hoof: therefore they are unclean to you' (Deuteronomy, 14:7). Zoroastrians and Copts are also not supposed to eat camels.

Ancient Greek writers refer to Persians banqueting on camel's meat. But it was the first-century Roman gourmet Apicius who popularized camel's foot as one of the supreme delicacies (together with such tempting dishes as tongues of peacocks and nightingales, roast ostrich and the brains of flamingos). Although Apicius compiled a cookery book full of recherché recipes, his recipe for camel does not seem to survive. In Rome grilled camel's feet became a choice dish. The decadent Roman emperor Elagabalus (r. 218–22), who came from Syria and who revered Apicius's writings, was particularly fond of camel's-foot stew.

The Arabist Geert Van Gelder comments on the comparative rarity of descriptions of cooking camel in early Arab poetry as follows: 'Bedouin dishes do not readily lend themselves to lyrical effusions, even though one may imagine how a starved Bedouin, who has lovingly described his she-camel, may not remain unmoved upon seeing a chunk of nourishing meat.'[19]

A seventh-century poet, Ibn al-Ahtam, described the preparation of a camel for the pot in the following unsparing terms:

Then I went to the sleeping line of couching beasts, where the camels great and fat in the hump, like towers, the best of their kind protected themselves from slaughter.

By putting in my way a fine white camel, wont to bring forth her young in the early spring, great as a stallion when she reared herself in front of the other camels ten months gone in calf.

I made for her with a sword-blow on the shank, or with a wide wound, spouting blood, in the stabbing place with its gash in front of the shoulders;

Then two butchers set to work upon her, and mounting atop of her they flay away her hide while yet she is breathing in the last gasps of death;

And there were drawn towards us her udder and her hump and a white camel calf that was just trying to stand, of purest breed,

Cut out of its mother belly: a brother bound to me by the brotherhood of the good, a true comrade, cleaves away from it with his sword the membrane that enwraps it.

And throughout the night, in the dark hours, there was set before us and the guest a roast of her flesh, rich with fat, and abundance to drink . . .[20]

If you are going to cook a camel, be aware of what you are taking on. In 2007 a French chef, Christian Falco, spit-roasted a camel. It took him fifteen hours and used three tons of wood and fifteen litres of oil; it fed five hundred.[21] In the desert a camel's meat is far too much for a single family, so they are only slaughtered when there is a celebration, like a wedding, that involves several households. Sometimes, after the camel has been slaughtered and cut up for meat, the fat in the hump is set aside and rendered into ghee and used for cooking later. The meat can be dried by being cut up in long strips and hung out to dry on bushes. Camels are slow to cook. A young unworked camel is best, as its meat is not so tough. Camel meat is low in fat and cholesterol, but high in iron and protein.

Udder, hump and calf are the most desirable meats. There seems to be a broad consensus that it tastes like beef.[22] But the *Larousse Gastronomique* compares a young camel's flesh to veal. I have eaten camel meat myself and would say that it tastes like venison. In Syria the fatty hump is judged to be so good that some people used to eat it raw. According to Dickson, 'The "hump" especially is considered a great delicacy in cold weather'. Though Dickson judged camel meat to be 'coarse and leathery', he admitted that the Bedouin love it.

Madame Guinaudeau, the author of a Moroccan cookery book, writing in the 1950s, described the arrival and slaughter of camels at Fez's Thursday market and how the meat then 'finishes up on the butcher's stall cut up in unappetising violet coloured pieces destined as mince meat for *kefta*'. Guinaudeau continues as follows: 'The white and sickly fat from the hump, cut out in huge thick petals decorates the stall and will be bought to make *khli* (preserved meat) of mutton, beef or camel. Less expensive, it can be cooked in the same way as beef, on a day when there is nothing else to be had, especially if the meat comes from a young camel'.[23]

For those who have not been put off by Guinaudeau, here is something a touch more flamboyant:

BAKED CAMEL (STUFFED)

500 dates
200 plover eggs
20 two-pound carp
4 bustards, cleaned and plucked
2 sheep
1 large camel
seasonings

Dig trench. Reduce inferno to hot coals, three feet in depth. Separately hard-cook eggs. Scale carp and stuff with shelled eggs and dates. Season bustards and stuff with stuffed carp. Stuff stuffed bustards into sheep and stuffed sheep into camel. Singe camel. Then wrap in leaves of doum palm and bury in pit. Bake two days. Serve with rice. Enjoy.

The above comes from the American novelist T. Coraghessan Boyle's novel *Water Music*, a novel about the explorer Mungo Park on an ill-fated expedition down the Niger River. In an interview, the author recommended horseradish to go with the meat.[24]

In *International Cuisine presented by California Home Economics Teachers*, you will find a vaguely similar recipe involving a medium-sized camel, a large lamb, twenty chickens, sixty eggs, twelve kilos of rice and so on. Allegedly it serves a 'friendly crowd of 80–100'.

Here is a more prosaic recipe from Jean-Marc Hervé of the Grand Union Pub, Westbourne Park, London:

1   Pan-fry 160g of diced camel meat per person with onions, garlic and herbs.
2   Add a pinch of flour.
3   Transfer to a deeper pan, add red wine and simmer.
4   Slow-cook for three hours.
5   Add sliced pineapple, chickpeas and curry paste. Heat until simmering, then spoon into pastry cases.
6   Serve with puy lentils, mashed potato and side salad.[25]

The fashion designer Stella McCartney denounced the Grand Union Pub for the barbarism of cooking camel. 'Lawrence of Arabia would be rolling in his grave if he knew about it.'[26] But Lawrence regularly ate camel, as is evidenced by this passage from *The Seven Pillars of Wisdom*: 'The last meal was good, partly because it was the last, partly because Mohammed had killed a suckling camel-calf for us and it had been boiled in sour milk by his wives who were famous cooks'.[27] Nevertheless, a TV chef who deserves to remain anonymous supported McCartney: 'In a civilized society, we should be encouraging people to conserve

A traditional Arab dish called the Quzi consisting of baby camel meat in the middle, roasted whole, surrounded by rice.

wildlife, not eat it into extinction.' But the pub gets its camel meat from the culling of feral camels in Australia. The feral camel population of Australia is actually expanding at an alarming rate and they are probably in less danger of extinction than the human race.

The taste of the milk depends upon what the camel has been eating, but it often tastes slightly salty. In East Africa, it is not just the nomads who consume the milk, but it is also sent to be sold in towns. Sometimes the milk is preserved by fermentation. Camel's milk has more fat, three times the vitamin c and a bit more protein than cow's milk. The sweetest milk comes from a camel that has just given birth. It is difficult, though not impossible, to make cheese or butter out of camel's milk as it does not coagulate easily. A sort of yogurt is also possible.

Finally with respect to dining on camel, the fourteenth-century North African historian Ibn Khaldun, looking back on the primitive, poverty-stricken life of the north Arabian tribe of Mudar in the pre-Islamic Hejaz, wrote that they 'were proud to eat *'ihliz*, that is, camel hair ground with stones, mixed with blood, and then cooked'.[28] However, he did not provide detailed guidance for the cooking.

4 Camels in the Medieval World of Islam

THE CAMEL-INFESTED IMAGINATION OF PRE-ISLAMIC ARABIA

Jamal is the common Arabic word for a mature male camel – that is, one in its seventh year and above.[1] *Jamil*, a word from the same trilateral root, means either melted fat or beautiful. For centuries Arabs have celebrated the beauty of the camel, even though at the same time some seemed to find something demonic in the creature. The old Arabian belief was that the camels were descended from the jinn and it was said that the resting places of camels were the haunts of jinn. According to a *hadith*, or saying concerning the Prophet Muhammad, he was once asked about the advisability of saying one's prayers on a spot where camels had rested. He replied 'Do not perform the prayers there, for they are the haunts of jinn.'[2] The association of camels with the demonic is also suggested by the fourteenth-century expert on dreams al-Damiri, who declared that to dream of a he-camel was to dream of the jinn, 'for they [male camels] are created out of the eyes of jinn'. According to an ancient Arab legend, the camels of Wabar, a pre-Islamic city in the Empty Quarter of Southern Arabia, mated with the camels of the jinn.

In poetry the camel was frequently compared to the ostrich and indeed the ostrich was classified as a kind of camel by the early Arabs. Apart from the similarity of its silhouette, the ostrich, like the camel, has flat feet and horny nails. In a fable recounted in Jahiz's *Kitab al-Hayawan* (Book of Beasts), the ostrich excuses itself from carrying a load on the grounds that it is a bird, but then excuses itself from flying by claiming that it is a flightless animal.[3] (In Persian *usturmurgh* means ostrich and in Turkish the word is *derekushu*, but both literally mean 'camel-bird', and in Latin the ostrich is *struthiocamelus camelus*.) The giraffe was seen by some as another type of camel, the result of a camel being crossed with a panther.

And, by the way, *jamal al-yahud*, literally 'the camel of the Jews', is Arabic for the chameleon.

In a treatise entitled *The Slaying of the Monks of Mount Sinai*, ascribed to St Nilus, there is a description of what appears to

John Frederick Lewis, *A Halt in the Desert*, 1855, watercolour.

have been camel sacrifice among the Arabs of the Sinai Desert in the late fourth century AD, in which a spotless white camel

> is bound upon a rude altar of stones piled together, and when the leader of the band has thrice led the worshippers round the altar in a solemn procession accompanied with chants, he inflicts the first wound . . . and in all haste drinks of the blood that gushes forth. Forthwith the whole company fall on the victim with their swords, hacking off pieces of the quivering flesh and devouring them raw with such wild haste, that in the short interval between the rise of the day star which marked the hour for the service to begin, and the disappearance of its rays before the rising sun, the entire camel, body and bones, skin, blood and entrails is wholly devoured.[4]

According to William Robertson Smith's famous *Lectures on the Religion of the Semites* (2nd edn 1894), this totemic form of ritual not only created a bond between the tribesmen and their god, but also increased the social solidarity of the tribe. Sigmund Freud in *Totem and Taboo* (1912–13) quoted St Nilus' account of camel sacrifice via Robertson Smith, but then argued that totemic feasting on the camel was a kind of symbolic re-enactment of the primal parricide. Subsequently however, Freud's *Totem and Taboo* was subjected to devastating criticism by the Romanian historian of religions and myths Mircea Eliade, who pointed out that as early as 1921 Karl Heussi had demonstrated in *Das Nilusproblem* that *The Slaying of the Monks of Mount Sinai* was not actually by St Nilus and that the unknown author of this work was drawing on the clichés of Hellenistic romances in its gruesome depiction of the behaviour of barbarians.[5]

Even so, it does seem to be the case in pagan Arabia that camels were ritually slaughtered at the time of the pilgrimage to a shrine full of idols at Mecca. Camels were also slaughtered in displays of wealth and hospitality. According to Lane's *Arabic-English Lexicon*, which here draws on the medieval Arab dictionary *Taj al-'Arus* ('Crown of the Bridegroom'), *'aqirahu* means 'He contended with him for superior glory and generosity in the hocking, or slaughtering of camels.' Lane adds that 'they did so for the sake of display and vain glory; wherefore the eating of the flesh of camels slaughtered on an occasion of this kind is forbidden in a tradition, and they are likened to animals sacrificed to that which is not God'. *Mu'aqara* is the noun meaning 'competition in the slaughtering of camels'.[6] An Arab's honour was established by this boastful kind of potlatch. The pre-Islamic aristocrat and poet Hatim al-Ta'i was famous for his extravagance in slaughtering hundreds of his father's camels in order to entertain visitors. His name became a byword for

An old photo of caparisoned camels at an Arab wedding in Suez.

A 3rd century BC Yemeni incense burner made of calcite, showing a camel and rider. The inscription, in Sabaean, names the pious donor.

generosity and as such he features in an *Arabian Nights* story.[7] A *jazur* is a camel that is set aside to be slaughtered. Camels were also sacrificed to the dead. A *baliyyah* is a she-camel which has been tethered at her master's grave and left to starve to death there.

Men often gambled over a slaughtered camel. *Maysir* was a pre-Islamic game in which the participants gambled for parts of a slaughtered camel.[8] Later it and all other forms of gambling were banned by the Prophet. *Maysir* literally means 'easy', or 'the game of the left-handed'. There were seven players in the game, seven win or lose arrows and three neutral arrows. Or maybe ten lots, in the form of these arrows, were drawn to determine who should win the best parts of the camel and who would pay for it. The men who drew the neutral arrows paid for the slaughtered camel. The udder and the hump were the desirable bits. The game was usually played by hungry men in times of hardship in winter. Although early students and anthologists of pre-Islamic poetry such as Ibn Qutayba (828–889) later tried to reconstruct how the game was played, nobody really knows, as after it was banned in the Quran the rules were forgotten.

Camel genealogies, which were traced in the female line commonly as far back as ten generations, were committed to tribal memory. The best camels were marked by having their ears slit. The standard bride-price was one hundred camels. This was also the blood-price paid in the case of a murder.

The camel loomed large in pre-Islamic poetry. As the great translator of early Arabic poetry, Sir Charles Lyall (1845–1920), wrote: 'the tending of camels pervades the whole of old Arabian poetry and words and metaphors drawn from it are in constant use for all manner of strange purposes.'[9] The *qasida*, or ode, normally began with the *nasib*, a lament for lost love and incorporating praise of that particular beast and sometimes also the poet's praise of himself for crossing the desert.[10] The camel section of the *qasida* is normally devoted to boasting (*fakhr*) about the beautiful and swift camel and/or the solitary and courageous rider. The *wasf* was the description of the camel (or sometimes the horse), and frequently the camel was compared to an antelope, ostrich or some other beast. Then came the *rahil*, the journey, usually alone on a she-camel. The she-camel, or *naqah*, was generally the favoured mount. The *qasida* normally finished with a *madih*, or panegyric, usually addressed to an actual or potential patron of the poet.

The *Mu'allaqa* of the sixth-century poet Imru' al-Qays is perhaps the most famous poem in the Arabic language. (*Mu'allaqa*, literally 'a suspended thing', refers to the tradition that a selection of seven of the best poems by various hands were hung up in a place of honour in the sacred enclosure in pre-Islamic Mecca). In the opening part of the poem, Imru' al-Qays pauses by the traces of an old campsite and remembers how in his youth, the preparation of his slaughtered camel for cooking was the prelude to his seduction of one of the women of the tribe. Here, in the somewhat archaic translation of A. J. Arberry, is the poet's description of the cutting up of his camel:

> Oh yes many a fine day I've dallied with the white ladies,
> and especially I call to mind a day in Dar Juljul,
> and the day I slaughtered for the virgins my riding beast
> (and oh, how marvellous was the dividing of its loaded
>   saddle),
> and the virgins went on tossing its hacked flesh about
> and the frilly fat like fringes of twisted silk.[11]

A possible implication of this passage is that the poet offered his camel to be slaughtered in exchange for sex. In what followed, Imru' al-Qays, older and sadder, mourning for lost youth, rode on (but on a horse).

The most famous eulogy of a camel was in Tarafa's *Mu'allaqa*, composed in the sixth century.[12] Here, again in Arberry's translation, is a part of Tarafa's description of his wonderful camel, which follows immediately after the poet's lamentation for lost love at a deserted campsite:

A late 12th–early 13th-century Kashan lustre vase in the shape of a camel.

Ah, but when grief assails me, straightaway I ride it off
mounted on my swift, lean-flanked camel night and day
    racing,
sure-footed like the planks of a litter, I urge her on
down the bright highway, that back of a striped mantle;
she vies with the noble, hot-paced she-camels, shank on
    shank
nimbly plying over a path many others have beaten . . .[13]

A pre-Islamic
Syrian statuette
of musicians on
a camel.

An early 17th-century Persian woven silk showing Layla with a camel train visiting her beloved Majnun in the wilderness.

And Tarafa goes on for another twenty-four lines about his camel's pasturage, her dry udders, the vertebrae of her neck and so on, ending up with her split upper lip and sensitive nose. Only then do we get an account of his actual journey. The great eighteenth-century Orientalist Sir William Jones commented on this poem that it 'it were to be wished that he had said more of his mistress and less of his camel.'[14]

Horses were more expensive and actually more useful in battle, but less exciting in the eyes of the poets. It was a common simile to compare war to a pregnant camel. The pre-Islamic poet Ta'abatta compared the thrusting of spears to camels drinking. In the fatalistic world of Pre-Islamic Arabia, the tribe might be compared to a herd of camels whose herdsman was death. The seventh-century poet Durayd fell in love with the poetess al-Khansa while watching her anointing a sick camel's scabs with pitch and this he celebrated in his poetry.

Dhat baww is a camel to which a *baww* is offered. A *baww* is the skin of her dead calf, stuffed and presented to her in order to deceive her into producing milk. (The practice continues to the present day.) Durayd once described his stricken plight in the following terms: 'I stood as a she-camel stands with fear in heart, and seeks the stuffed skin with eager mouth, and thinks – is her youngling slain?'

THE CAMEL'S LOVE OF POETRY

Camels respond to poetry as well as music. *Tirijiza* is Arabic for 'he urged, or excited, his camels, by singing *rajaz*'. *Rajaz* is the noun commonly used to describe a certain sort of simple metre (an irregular iambic metre consisting of four feet to the line: - - u -) used in the simple songs by labouring folk engaged in rhythmical activities such as fishermen, boatmen, water-carriers and camel drivers. However the proper original meaning of the word is 'tremor, spasm, convulsion (as may occur in the behind of a camel when it wants to rise)'.[15] To return to the more poetic meaning, in 1999 the progressive rock band Camel issued a recording entitled *Rajaz*. According to the blurb accompanying the CD: 'The music of poets once carried caravans across the great deserts. Sung to a simple metre of the animal's footsteps, it transfixed weary travellers on their objective . . . journey's end. This poetry is called "rajaz". It is the rhythm of the camel.' The album is notionally set in the ancient Near East.

THE LANGUAGE OF CAMEL

The Arabs developed a fabulously rich vocabulary of camel-related words. First, as the explorer Sir Richard Burton observed, 'There is a regular language to camels. "Ikh! Ikh!"

makes them kneel; "Yáhh! Yáhh!" urges them on; "Hai! Hai!" induces caution, and so on.'[16] To which one might add 'Hut! Hut!', which is used to urge the camel forward.

The approach of early Islamic scholars to the camel was chiefly philological. There is an old joke that every Arabic verbal root has four meanings: first its primary meaning; then a meaning that is the opposite of the first; then a camel-related meaning; and finally a sexual meaning. Indeed, the Arabs have an enormous number of words for different kinds of camel. Some have already been mentioned. Joseph Freiherr von Hammer-Purgstall, a great eighteenth-century Orientalist, collected 5,774 words for camel and camel-related features and paraphernalia. Many of the terms he collected were poetical metaphors.[17] But, for example, there really are at least thirty different words for camel milk.

The desert explorer Wilfred Thesiger was impressed by the camel-related vocabulary of the Bedouin of Arabia's Empty Quarter:

> To talk intelligently to the Bedu about camels I tried to learn the different terms which they used, and these, numerous enough in any case, tended to vary among the different tribes. They used several different words for the singular and the plural. They had different names for the different breeds and colours, for riding camels and herd camels, and a different term, which varied according to the animal's sex, for a camel in each year of its life until it was fully grown, and others for as soon as it began to grow old. They had terms for a barren female, and for one in milk, which varied again depending how long she had been in calf or in milk. I listed many of these words but found it impossible to carry most of them in my head.[18]

Here are some Arabic words. Strictly speaking *jamal*, the
common Arabic word for a camel, applies only to a male camel
between its sixth and twentieth year. The word, when it appears
in literature, lacks the poetic, emotional and metaphoric reso-
nance of *naqa*. A *naqa* is a fully grown she-camel from its fifth
year onwards which is used for riding. The *naqa* is, as we have
seen one of the regular stars of the *qasida*. *Ibl* is the common
collective term for camels. A *dhalul* is a riding camel regardless
of sex. *Ayhama* is a strong she-camel. A *talih* is a she camel that
is tired or emaciated. *'Awamil* is a camel with two humps. A
mudhkir is a camel that only gives birth to male foals (and this
is unlucky). A *sarma* is a she-camel whose milk supply has

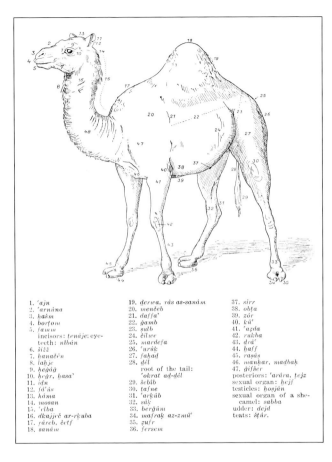

Alois Musil, the 46 parts of a camel in Arabic, from *The Manners and Customs of the Rwala Bedouin . . .* (1928).

1. *ʿajn*
2. *ʿarnùna*
3. *hašm*
4. *borṭom*
5. *famm*
   incisors: *tenáje*; eye-
   teeth: *níbán*
6. *šiźź*
7. *hanaćén*
8. *lahje*
9. *heǵàǵ*
10. *heǵr, hasaʾ*
11. *idn*
12. *fáʾ ùs*
13. *háma*
14. *mosan*
15. *ʿelba*
16. *dkajjćč ar-rkuba*
17. *ráreb, ćetf*
18. *sanám*
19. *derwa, ràs as-sanàm*
20. *menćeb*
21. *daffaʾ*
22. *ǵamb*
23. *sulb*
24. *ćilwe*
25. *mardefa*
26. *ʿurùk*
27. *fahad*
28. *ḍél*
    root of the tail:
    ʿokrat ad-dél
29. *šebíb*
30. *tafne*
31. *ʿarkùb*
32. *sák*
33. *berǵám*
34. *mafrak az-zmùʿ*
35. *zufr*
36. *fersem*
37. *sirr*
38. *obṭa*
39. *zór*
40. *káʿ*
41. *ʿazda*
42. *rukba*
43. *dráʾ*
44. *haff*
45. *rasàs*
46. *manhar, madbah*
47. *ǵifšer*
    posteriors: *ʿaràra, ṭejz*
    sexual organ: *hejf*
    testicles: *hosjàn*
    sexual organ of a she-
    camel: *sabba*
    udder: *dejd*
    teats: *šṭàr.*

been cut off. A *hadin* is a man who urges or excites camels by singing to them.

So much for fairly ordinary camel vocabulary in Arabic. Here are some more recondite terms (given mostly in the slightly archaic diction of the nineteenth-century lexicographer Edward William Lane):

A 19th-century calligraphy print from northern India.

Mabudun: a camel having the pastern of his foreleg tied or bound to his arm, so that his foreleg is raised from the ground.

Abil: a camel content or satisfied with green pasture so that there is no need of water.

Mathurun: a camel having a mark made upon the bottom of his foot with an instrument called a *mithra* in order that his footprints on the ground may be known.

Arifun: a camel suffering pain in its nose.

Bahiratun: a female camel having her ear slit of which the mother had brought forth ten females consecutively before her, and of which the ear was slit.

Ajab: a camel with its hump cut off.

Majrub: a man who has his camels affected with mange (as in the proverb 'There is no god to the man whose camels are affected with mange').

Jalad: the skin of a young camel, which (being stripped off) is put over the body of another camel, in order that the mother of the skinned one may conceive an affection for it and suckle it.

Ahjal al-ba'ayr: he loosed the camel's shackle from his left fore-leg and fastened it upon the right. (One medieval Arabic dictionary has the procedure the other way around.)

Detail from 'An Arab Dwelling in the Desert, Suez', lithograph after a drawing by Manó Andrásy, a Hungarian who toured the Levant in the early 1850s.

Bihi hazzun: said of a camel, he has an incision, or cut, in the edge of the callous protuberance upon his breast produced by his elbow, which makes it bleed. (There is another Arabic word for the same thing but without bleeding.)

Muhataba: a she-camel that eats dried thorns.

Rugha: the cry or grumble of a camel.

Argha: he made his she-camel to utter the grumbling cry termed *rugha* and he made his camel do so in order that he might be entertained as a guest.

Khursi: a she-camel that does not utter the cry called *rugha*.

Farij': a she-camel that has come apart in parturition and hence hates the stallion and dislikes his being near.

Khazuqun: a she-camel that pierces the ground with her toe.

Khatirun: the flailing of a camel's tail between the parts above his thighs, when he moves about.

Makhalif: camels that have been pastured on fresh herbs or leguminous plants, and have not fed on dry herbage, and to which their pasturing of the former has been of no avail.

Khaliqunu: a she-camel that is made to affect, with another she-camel, one young one, so that both yield their milk to it, and to which the people of that tent or house, confine themselves exclusively, of the other for the purpose of milking her.

Khimsun: the drinking of a camel on the fifth day, counting the day of the next preceding as the first; their drinking one day, then pasturing three days, then coming to water on the fifth day, the first and last days on which they drink being thus reckoned.[19]

A lexical item not found in Lane's unfinished dictionary is *ista-wanaqa*, 'he mistook male camels for female camels' (which is as much as to say that he is a complete clot).

A 16th-century Bukharan gouache miniature of a composite camel and attendant.

Léon Belly, *Pilgrims Going to Mecca*, 1861, oil on canvas.

In Persian a *nakhur* is a camel who needs her nostrils tickled before she will give milk. *Nar* is Kirghiz for a tall Bokharan camel. As in the traditional Kirghiz saying is 'One never falls but from a *nar*'.

KORANIC CAMEL

It seems likely that most of the camel-related vocabulary was elaborated in pre-Islamic times, that is to say prior to the 620s. But it was an almost entirely oral camel culture and it was only after the coming of Islam that much was written down about the camel and works of prose were devoted to it. In a lecture given in Buenos Aires in 1951, 'The Argentine Writer and Tradition', Jorge Luis Borges made the following assertions:

A few days ago, I discovered a curious confirmation of the way in which what is truly native can and often does dispense with local colour; I found this confirmation in Gibbon's *Decline and Fall of the Roman Empire*. Gibbon observes that in the Arab book *par excellence*, the Koran, there are no camels; I believe that if there were any doubts as to the authenticity of the Koran, this lack of camels would suffice to prove that it is Arab. It was written by Mohammed, and Mohammad, as an Arab, had no reason to know that camels were particularly Arab; they were for him a part of reality, and he had no reason to single them out, while the first thing a forger, a tourist, or an Arab nationalist would do is bring on the camels, whole caravans of camels on every page: but Mohammed, as an Arab, was unconcerned; he knew he could be Arab without camels.[20]

There are several things wrong with the above. Mohammed was illiterate and he did not write the Qur'an. It was compiled from his oral revelations after his death. Secondly, Gibbon made no such claim about the Qur'an, but one of his footnotes seems to have been misread by Borges. Thirdly, there are many references to camels in the Qur'an.

The Qur'an shows far more interest in animals than does the Bible and camels feature particularly prominently. Just some of the relevant passages will be cited here:

What do they not consider how the camel was created,
How heaven was lifted up,
How the mountains were hoisted,
How the earth was outstretched?
Then remind them! Thou art only a reminder . . .
(Qur'an 88:17–21)

The sura or chapter entitled 'Hood' tells the story of Salih, one of the pre-Islamic prophets who had been sent by God to warn the people of Thamud to amend their ways, to cease worshipping pagan divinities and repent. But they doubted that he was a prophet and demanded a sign from God, whereupon the She-camel emerged from a rock. Salih said:

> 'O my people, this the She-camel of
> God, to be a sign for you. Leave her that
> She may eat in God's earth, and touch her
> not with evil, lest you be seized by a
> nigh chastisement'.
> But they hamstrung her; and he said,
> 'Take your joy in your habitation
> Three days—that is a promise not
> to be belied.' (Qur'an 11:64–70)

But the people of Thamud were unimpressed by Salih and the miracle and they hamstrung the camel and then killed it. After three days the Cry of God destroyed the unbelievers of Thamud, sparing only Salih and those who believed with him. The Qur'an also partly echoes the New Testament, specifically Matthew 19:24: 'It is easier for a camel to go through the eye of a needle, than for a rich man to enter the kingdom of God.' Whereas the Qur'an has:

> Those that cry lies to Our signs and wax proud against them –
> The gates of heaven shall not be opened
> To them, nor shall they enter Paradise
> Until the camel passes through the eye
> Of the needle. (Qur'an 7:39)

An 18th-century Iranian illustration from the *Qisas al-Anbiya*, a compendium of Koranic stories, showing the wise man Salih and a she-camel emerging miraculously out of a rock.

TRADITIONAL CAMEL

According to tradition, the Prophet reprimanded people who sat idly on the backs of their camels in the marketplace, using them as chairs. He said 'Either ride them, or leave them alone.' A Bedouin asked the Prophet whether there would be camels in paradise and he was told that 'Everything one longs for will be there.' Another saying of the Prophet teaches that one should 'Speak ill neither of the camel nor of the wind; the camel is a benefit to man and the

89

wind is an emanation from God.' According to another *hadith*, or saying of the Prophet, 'Camels are a glory to their people, goats and sheep are a blessing and prosperity is tied to the forelocks of horses to the Day of Judgement.' Again the Prophet compared camels to the Qur'an, for they both need careful attention. On the other hand, he is also on record as saying that 'camels are the offspring of devils'.

THE BATTLE OF THE CAMEL

This battle, which took place near Basra in 656, was fought between followers of Ali, the cousin and son-in-law of the Prophet who claimed the caliphate, that is the leadership of the Muslim community, and the followers of Aisha, the widow of Muhammad, who supported the claims of al-Zubayr, a well respected early convert to Islam, to the caliphate. This was the first battle in which Muslim fought Muslim. It is called the Battle of the Camel because Aisha presided over her side of the battle on a scarlet basketwork palanquin on the back of a camel. Her camel was called Askar and wore a coat of mail. As Barnaby Rogerson has described the scene:

> In the late afternoon light the fighting concentrated round the camel litter in which Aisha sat, protected by armoured panels. Champion after champion from her ranks came forth to take the place of honour – and near certain death – by advancing to hold the camel halter of the Mother of the Faithful and serve as her protective knight. One by one they were felled by the surrounding archers, as if an ode from some old pre-Islamic battle verse were being brought back to life.

According to al-Damiri, it is said that the hands of eighty men were cut off while they held the halter of the camel. Eventually it was hamstrung and Aisha was brought low. Victory went to Ali and she surrendered. The defeated were generously treated.[21]

The camel became an icon of the camel-rearing, Muslim Arabs. In the early years of the Islamic conquests Basra in southern Iraq was established in the first instance as a garrison encampment. But it soon grew into a great city that was a centre for literature, science and philosophy. The Mirbad was the chief commercial and intellectual centre. Mirbad means literally a place for tethering animals and that was indeed one function of the great open space, as the Bedouin parked their camels there. Arabic was the language of the Qur'anic revelation. Moreover, all of a sudden Arabic had become the language of a

A Persian Mina'i fritware jug of c. 1200 showing a camel train.

great empire. But the language was a difficult one and there were many contentious points regarding the vocabulary and grammar of the Qur'an and early Islamic poetry. The desert-dwelling Arabs were known to be the guardians of the purest form of Arabic and their memories were the best guide to how things had been said and done in the lifetime of the Prophet. So it was that theologians, grammarians and poets living in Basra went out regularly to the Mirbad to talk with the camel-rearing men from the desert and so it was that refined urban poets learned to sing about deserted campsites, the beauties of camels and the rigours of desert life.

The great philologist Al-Asma'i (740–828?) haunted the Mirbad and made a particular study of Bedouin vocabulary. One outcome of his study was the *Kitab al-Ibl* ('The Book of the Camel'), a study of camel-related words in pre-Islamic poetry. Though he had collected most of this vocabulary by talking to Bedouin who had ridden in from the desert to trade in Basra, he also rode out into the desert to collect more obscure words. He regarded the simple ways of the desert-dwelling Bedouin as the ideal form of life.[22]

Al-Jahiz (*c.* 776–868), a Basran and the Arab world's greatest ever essayist, produced an enormous treatise, the *Kitab al-Hayawan* ('Book of Beasts').[23] While he drew heavily on Greek and other literary sources, he also got a lot of his information from Bedouin informants. Sadly he never got around to writing the section on the camel, and he has more to say about pigeons than camels. Perhaps he was avoiding the subject because so much had already been written on it, or perhaps the subject was so complex and important that he was leaving it until last. Nevertheless, scattered throughout his magnum opus, one can find a fair bit of information about camels. For example, he refers to the Jahili, or pre-Islamic custom that when a herd reached one

An 18th-century illustration for the 'Story of the Camel-Driver and the Snake' from the *Kalila wa Dimna*, a Persian and Arabic compendium of animal fables.

thousand, the thousandth camel had an eye torn out and when that number was exceeded the second eye was torn out. Such an animal was known as *mu'amma* (the blinded one) and celebrated in poetry.[24] (Enucleation is the technical term for the tearing out of an eye.) Al-Jahiz noted that camels only like to drink from stagnant water. He took the trouble to refute the idea that the giraffe is the offspring of the coupling of a camel with a tigress. He also related the legend that the serpent once had the form of a camel, but God punished it by forcing it to crawl on its belly on the earth.

Al-Sahib ibn 'Abbad (938–95) began his career as a poet in Basra but became a vizier of the Buyid princes in tenth-century Persia. He was a great patron of literature and allegedly possessed a library of 117,000 volumes. It is said that on his many travels as a warrior and statesman he never parted with his beloved books. They were carried about by four hundred camels trained to walk in alphabetical order. His camel-driver librarians could put their hands instantly on any book their master asked for. Sadly, this seems to be apocryphal, but the germ of it is an anecdote that is very likely true in which Ibn 'Abbad refused to become the vizier of a Samanid ruler on the grounds that he would have to travel and that it would be too expensive to move his library on the backs of four hundred camels.[25]

In Basra, some time in the tenth or eleventh centuries, a secretive group known as the Ikhwan al-Safa' (Brethren of Purity) produced an encyclopaedia entitled *Rasa'il* (Letters), covering all the sciences from a philosophical and moral point of view. Their encyclopaedia includes a curious fable in which the animals, birds and insects bring a court case against humanity, charging the species with cruelty, ecological heedlessness and greed. The case is brought before the court of the King of the Jinn and the camel is one of the witnesses for the prosecution. After man has put the case that animals have been made the servants of man, the jackal, leading the prosecution, responds and argues against this, then the ass and the ram complain about the way they are treated by the Adamites (mankind).

The camel joined in, 'Also had you seen us your Majesty, as prisoners of the Adamites, our noses bound up with rope and our halters in the hands of drivers who forced us to carry heavy loads and make our way in dead of night through dark defiles and waterless plains over a rocky track, bumping into boulders and stumbling with our tender pads over rocks and rough and broken ground, hungry and thirsty, our sides and backs ulcerated and sore from the rubbing of the saddles, you would have pitied us and wept for us. Where then is their pity?'

The elephant, the horse and the mule take up the complaint and the case continues, but in the end the animals lose.[26]

A similar complaint about the cruelty of men appears in the *Arabian Nights* story of 'The Peahen, the Duck and the Gazelle', in which a bold lion cub is warned by a succession of beasts about the cunning and oppression of the 'son of Adam'. Again, somewhat similarly, the misanthropic blind vegetarian poet, Abu al-'Ala al-Ma'arri (973–1058) in his treatise *Risalat al-sabil wa'l-shahij* (Epistle of the Neigher and the Brayer) has the camel complain of its hard life under the despotic rule of mankind:

One of the strange acts of humans is that, when they want to travel in a country without water, they deprive camels of water for eight days, then, when they are all but exhausted from thirst, they let them drink their fill. Then they go out into the desert: and if water is scarce, they cut their bellies open and drink the liquid contained in it . . . they drink our blood, which they let in time of drought . . . no animal has to suffer from humans what camels have to suffer: they exhaust them while travelling, and feed them in the desert to beasts and birds of prey . . . Juwayriyya Ibn Asma', from the tribe of Fazara boasts of having slaughtered his mount for a wolf that he met in the

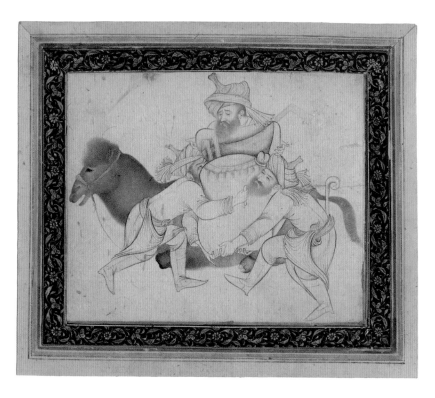

Men loading a
camel on an
Iranian album leaf,
early 17th century.

desert . . . Has not a camel more rights than a wolf? Does
not somebody who is close deserve more respect than some
stranger? Had his guest been a human being, someone of
his own kind, he would at least have had an excuse.[27]

Al-Ma'arri also wrote a poem in which the passage of time is
compared to a caravan of camels:

> Meseemeth the days are dromedaries lean and jaded
> That bear on their backs humanity travelling onward;

They shrink not in dread from any portentous nightmare,
Nor quail at the noise of shouting and rush of panic,
But journey along for ever with those they carry,
Until at the last they kneel by the dug-out houses.[28]

Al-Damiri, a religious scholar in fourteenth-century Cairo, wrote a zoological treatise, *Hayat al-Haywan* ('Lives of the Animals'), in which camel lore features twice, in articles devoted to *Ibl* and *Jamal*. (*Ibl* refers to a plurality of camels, while *jamal* refers to an individual male camel.)

According to al-Damiri, camels are metaphorically known as *banat al-layl* (daughters of the night). He also offers some characteristically naïve dream divination: 'He who dreams of acquiring a hundred camels will rule over a body of influential men and will acquire a large fortune.' Also 'He who eats the flesh of a camel in a dream will fall ill.' Dreams about he-camels are dreams about the jinn, because they are created out of the eyes of the jinn.[29] (According to Richard Burton, 'The camel carries the Badawi's corpse to the cemetery which is often distant; hence to dream of a camel is an omen of death.'[30]) Al-Damiri also drew attention to the *jamal-bahr*, a large fish in the shape of a camel. According to Islamic law, it is permissible to eat it.[31]

Camels and camel-riding were closely associated with Arabism by the people that the Arabs had conquered. An early ninth-century Persian noble declared that 'I have become one of these people in everything that I detest even unto the eating of oil, and the riding of camels, and the wearing of sandals.'[32]

THE MINIATURE CAMEL

The camel featured only occasionally in medieval Arab book illustration, but there are plenty of camels in Persian and Turkish

A 13th-century tile from the Takht-i Sulayman, a palace in Azerbaijan, shows a scene from the *Shah-nama* in which the Persian king, Bahram Gur, shoots deer from a camel. His harpist rides on the same animal.

A miniature of camels by al-Wasiti, from the *Maqamat al-Hariri.* [Yahya ibn Mahmud al-Wasiti (13th century) Abu Muhammad al Qasim ibn Ali al-Hariri (1054–1122)].

A drawing by
Mu'in Musavvir
of Isfahan in Persia
(*fl.* 1635–97) of the
rear view of a
camel.

miniatures. Occasionally the Prophet was shown by Persian miniaturists seated on a camel, presiding over the Day of Judgment.
Camels fighting was a popular, even stereotypical, subject for
Persian miniaturists.[33] So too was the legendary prowess of
Bahram Gur, legendary Persian prince and huntsman. The
story goes that he was riding out on a dromedary with Azda, his
favourite slave girl, seated behind him. She challenged him to
shoot arrows in such a way as to turn a female deer into a male,
to turn a male into a female, and to pin a gazelle's ear to its
leg. Bahram shot two arrows into the head of a female, thereby
giving her horns. He used two arrows to shoot off the horns of a
buck and he shot a deer through the ear and foot with the same

arrow. But though he succeeded, she sneered that he must be a devil to have performed such feats. Enraged, he threw her off the camel and used it to trample her to death. The image of the pair on their camel, besides appearing illustrated manuscripts, was a common motif on glazed Persian tiles.

5 The Beauty of the Beast: Literature and Art

It is unfortunate that the only study of the camel in literature is B. Baast's *Gangan ulaan temee* (Ulan Bator, 1975). Since this book is in Mongolian, I have no idea what it says. Turning now to Japan, before any camels were seen in that country, the Japanese read about Bactrians in Chinese literature and fantasized about them. Indeed the camel had the same sort of status in Japan as the unicorn and the manticore did in medieval Christendom. References to the camel featured prominently in Japanese erotica. The Japanese for camels is '*rakuda*', and this sounds like '*raku da*!', meaning 'feel comfy!' Therefore the camel became the emblem of male–female conviviality.[1] As Tetsuo Nishio observes: 'A sexual folklore developed concerning camels, and this folklore, together with romantic images of desert life, gave rise to the typical fantasy about a camel-riding prince and princess in bridal procession. This scene is still fostered in the hearts of almost all Japanese by way of the famous nostalgic song called "The Dune in the Moon".'[2] The Dutch East India Company was the first to import a pair of dromedaries into Japan in 1821 and they were paraded around the country for people to marvel at. Tani Buncho, a well-known painter of the late Edo period, produced a fairly realistic painting of the pair.

In China ceramic figures of the camel (in Chinese, *luotuo*) were produced in great numbers to furnish graves. The practice began in the Han period (206 BC–AD 220) and continued thereafter for a thousand years. Prior to the Han period camels were mainly harnessed to carts, but with the opening up of the Silk Route,

Kuniyasu Utagawa,
A Camel, 1824,
woodblock print.

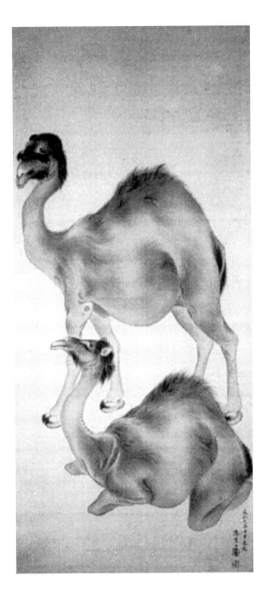

Maruyama Oshin,
Camels, 1824,
colour and ink on
silk hanging scroll.

Wu Zuoren
(1908–97),
Camel Herding,
1977, ink and
colour on paper.

Bactrians were more commonly used as pack animals and that is what the statuettes show. (The Chinese did not ride camels, even though some statuettes show foreigners or musicians sitting on them.) Since the Bactrian camel was used to transport luxury goods, it became a symbol of prosperity and hence an appropriate creature to put in a rich man's tomb. As the centuries passed, representations of the camel became less realistic and more loaded with symbolism. Later figurines often show large masks hanging on the sides of the camel. Elfriede Knauer, who has made a special study of these statuettes, suggests that masks may indeed have been attached to the saddles in order to scare the goblins of the desert. The popularity of camel statuettes peaked in the Tang period (618–907). From the thirteenth

century onwards statuettes of funerary camels ceased to be manufactured. Chinese painters and potters depicted both the dromedary and the Bactrian and it seems that some sixteenth-century artists thought that the difference between one- and two-humped camels was that of sex.[3] Incidentally 'qilin' is

Hua Yan, *Deep Snow on Tianshan*, 1755, hanging scroll, ink and colour on paper.

Chinese for 'camelopard' and this also features in Chinese iconography. The qilin, which is often shown with a dragon's head and hooves, walks on grass without disturbing the blades of grass. The Japanese version of the qilin is kirin. The kirin more closely resembles a giraffe and it has given its name to a brand of Japanese beer.

MEDIEVAL CAMEL

The *Roman de Renart*, compiled in the late twelfth and early thirteenth centuries, is a collection of tales featuring the wily and outrageous behaviour of Renart, a fox in the kingdom of animals. In what is known as Branch Va of the *Roman*, featuring the lawsuits of Isengrin the Wolf and Bruin the Bear, two camels appear. Isengrin is pursuing a case against Renart at the court of the lion king. The king needs advice: 'Beside the king sat the camel, a great favourite at court. He came from Lombardy to bring my lord Noble tribute from Constantinople. He been sent by the pope as his legate and friend, and he was very wise and a good jurist.' Notwithstanding the last statement the camel, when asked if by the king if he has ever come across a parallel case, proceeds to spout a great deal of pompous nonsense sprinkled with pointless Latinisms and Italian expressions.[4] A second camel called Musart features in the same story as one of the beasts who rallied to Isengrim's revolt. Although the camels have what are not much more than walk-on roles in *Le Roman de Renart*, this was probably their first appearance in Western fiction. The camel and the lion were totally exotic fantasy animals for French readers. Therefore it was easy to land them with fantastic roles unlike the more familiar fox and rabbit. The anonymous contributors to the various versions of the *Roman* enjoyed fooling around with scale. And so, for example, the

improbable fantasy of a hedgehog beheading a camel was conjured up by them.

In the *Roman de Renart* the camel did not stand for or sym-
bolize any abstract quality. But elsewhere in medieval literature,
the camel was assigned more symbolic meanings than it could
comfortably carry. It variously stood for humility, wealth,
greed, or lust. The kneeling camel was often carved on miseri-
cords and bench-ends as a symbol of humility.

John Ruskin in his *The Bible of Amiens* (1885), a study of the cathedral and especially the medieval sculptures of its west porch, touched on the camel:

> OBEDIENCE, bears shield with camel. Actually the most disobedient and ill-tempered of all serviceable beasts – yet passing his life in most painful service. I do not know how far his character was understood by the northern sculptor; but I believe he is taken as a type of burden-bearing, without joy or sympathy, such as the horse has, and without the power of offence, such as the ox has. His bite is bad enough (see Mr Palgrave's account of him) but presumably little known of at Amiens, even by Crusaders, who would always ride their own horses or nothing.[5]

In 1133 or 1134 a ceremonial robe of red silk embroidered with gold thread was made for Roger II, the first Norman King of Sicily. The symmetrically devised design showed two ad-dorsed lions savagely leaping on the backs of two camels. It is thought that this image symbolizes the triumph in Sicily of

A camel with a monkey on its back, from a 13th–14th-century Latin bestiary.

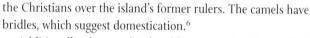

the Christians over the island's former rulers. The camels have bridles, which suggest domestication.[6]

Additionally, the camel in heraldry conventionally symbolizes docility, patience and perseverance. (It has also featured as a supporter of heraldic arms, for example those of Lord Kitchener.) Beryl Rowland in *Animals with Human Faces: A Guide to Animal Symbolism* (1974), after citing medieval instances of the camel as an image of humility, prudence or stupidity, concludes as follows: 'But the camel's most conspicuous role was sexual: it was a medieval nymphomaniac.'[7] This perhaps echoes Jeremiah 2:23, where Israel was denounced for whoring after false gods like camel on heat: '*thou art* a swift dromedary traversing her ways'. The camel in the Western Middle Ages was above all famed for its voracious sexual appetite and its rutting violence. In *The Canterbury Tales*, in the tailpiece to the 'Tale of the Clerk of Oxenford', Chaucer urged women to 'be strong as is a greet camaille;/ Ne suffreth nat that men yow doon offense'.[8]

In medieval and renaissance England the camel's hump was an emblem of the rich man's superfluities and it is probably because of this assocation that Edmund Spenser in *The Faerie*

Queene made the camel the image of Avarice in his pageant of the Seven Deadly Sins.[9] Though, of course his readers would also have thought of the New Testament: 'It is easier for a camel to pass through the eye of a needle than for a rich man to enter the kingdom of heaven.'

There was a lot of speculation in medieval Christendom about the camelopard. This monstrous creature had the body of an ass and the head of a camel. Alternatively, the camelopard was a half camel and a half leopard, a product then of an improbable miscegenation. The conjuring up of the camelopard may have been based on distant sightings of a giraffe. The fantastic zoology of the Middle Ages also had niches for the allocamelus (the body of a camel with the head of an ass) and the camelopardel (the same but with two horns curved backwards). Fourteenth-century writers also speculated about the camelion, which was the alleged offspring of a camel and a lion.

Heraldic 'escutcheons' (shields) including camels.

GREAT ARTISTS AND THE CAMEL

From the late Middle Ages onwards, the camel featured regularly in paintings of the Adoration of the Magi. The first time that camels made their appearance in this scene was possibly in Giotto's version, painted for the Scrovegni Chapel in Padua in 1303–5. The camels in question are not very impressive. They look like long-necked donkeys with whiskers. The camel also made an early appearance in the version of the *Adoration of the Magi* painted for the *Très Riches Heures du Duc du Berry* around 1410. In Gentile da Fabriano's version of the same scene, painted around 1423, the early hours of the baby Jesus seem to have spent in a regular zoo, as, besides the camel, the painting includes a greyhound, a lioness, a leopard, two apes, three falcons, a dove, an ox and an ass. In Renaissance art, the

Pisanello's 1430s
front view of a
camel.

camel functioned simultaneously as a marker of the exotic and of the authentically biblical. The staging of camel fights, which took place in Medici Florence, may have provided some of the artists with their models.

In emblem books the camel featured as the emblem of Asia. But when Giovanni Battista Tiepolo (1696–1776) came to do his

Giotto, *The Adoration of the Magi* including a camel, 1304–6, fresco in the Capella degli Scrovegni, Padua.

Giambattista Tiepolo, Asia emblematised by a camel in a 1751 fresco of the four continents in the Prince-Bishop's Palace ('Residenz'), Wurzburg, Austria.

great frescoes of the four continents in Wurzburg, he perversely portrayed the personification of Africa as seated on a camel, whereas Asia was shown as riding an elephant.[10] Tiepolo seems to have been fond of depicting camels. Vulpine-looking camels feature in his painting *Rachel Hiding the Idols from her Father, Laban*, and there is also a curious drawing of Pulcinello arriving in Egypt on a camel. Camels were among the exotic furnishings of Tiepolo's imaginative world.

Curiously, the absence of camels has also featured prominently in Western art. In particular the absence of camels in Nicolas Poussin's painting of *Eliezer and Rebecca* (1648) has attracted academic attention. In the biblical story (Genesis 24) the servant Eliezer was sent by Abraham to find the right wife

for his son Isaac. As Eliezer neared the city of Nahor, he resolved that the first woman to offer him and his ten camels water to drink would be the right wife for Isaac. Then at a well a woman offered him a drink. 'And when she had done giving him drink, she said, I will draw water for thy camels also, until they have done drinking.' That woman was Rebecca. Yet in Poussin's painting of the scene (now in the Louvre), Rebecca is shown offering Eliezer water from the well, but the camels are nowhere in sight. What is going on? Might it be that the camels are not there because Poussin was no good at painting camels? This is not in fact the case, since he did two other versions of *Eliezer and Rebecca at the Well*, one before and one after the Louvre painting, and in both these paintings the camels feature, though admittedly not the full complement of ten. Poussin was capable of painting fine camels.

In 1668, during a session of the Académie Royale de Peinture et de Sculpture, Philippe de Champaigne debated the issue with Charles Le Brun. Champaigne insisted that camels should have been included, as their inclusion would have made the scene earthier and less idealized and their ugliness would have set off the beauty of the women. Besides, Poussin had betrayed the Holy Writ, and the absence of the camels made it harder to identify which incident in the Bible was being portrayed. Le Brun however asserted that camels were such ugly or bizarre creatures that they would have destroyed the classical serenity of Poussin's picture. Camels should never feature in serious painting. Le Brun insisted that camels lacked *bienséance*. (It is distressing that both the disputants took the ugliness of camels for granted.) Le Brun also thought that that the presence of camels in the painting might have led the viewer to confuse Eliezer with some long-distance travelling merchant. Then again, other later scholars have argued that Eliezer's selection of

Rebecca may be seen as prefiguring Gabriel's Annunciation to the Virgin Mary and camels would certainly be out of place in a lightly disguised Annunciation scene. It should be borne in mind that Rebecca's task would not have been a light one and not just a matter of presenting a jug of water or two to the beasts. Ten camels would easily have been able to drink two hundred

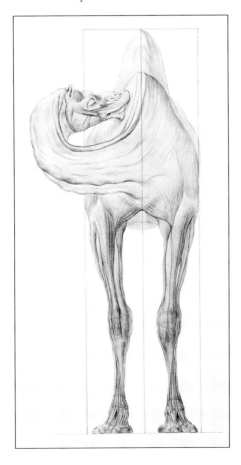

An end-view of a camel in Elijah Walton's book *The Camel* (1865).

gallons. Perhaps Poussin was aware of this and did not want to paint such a scene of heavy labour.[11]

In the course of the nineteenth century the Middle East was opened up to Western colonialists, tourists and artists. Elijah Walton (1832–1880) was a Victorian painter of Alpine landscapes who turned himself into an expert on camels. After a long time spent in a Bedouin encampment near Cairo, he produced *The Camel: Its Anatomy, Proportions and Paces* (London, 1865), with ninety-four meticulously drawn plates, some of them in colour.[12] More generally, camels of course appeared in nineteenth-century Orientalist painting. For example, when Horace Vernet, a specialist in biblical themes, painted *Judah and Tamar* (1840), he showed Judah leaning lecherously over Tamar, who covers her face but shows a lot of leg, and behind them and towering over the pair stands a handsome saddled dromedary. A camel is the chief subject of Thomas Seddon's *Dromedary and Arabs at the City of the Dead, Cairo with the Tomb of Sultan El Barkook in the Background* (1856). A host of camels appear in Léon Belly's great painting *Pilgrims Going to Mecca* (1861). Belly depicted the camels and their riders progressing through a desert landscape that has been bleached of colour by the intensity of the sun. In *The Orientalists: Western Artists in Arabia Persia and India*, Kristian Davies has this to say of the painting: 'One can feel that sensation in *Pilgrims*, that the animals are not walking but hovering over the landscape like a mirage, that the caravan is actually a great floating ship somehow moving over the sand like an ocean raft.'[13]

But though plenty of other examples of the camel in Orientalist painting can easily be found, they do not feature as frequently as one might have expected. This is because the Victorian art-buying public was mostly mad on horses. The Orientalist painter Eugène Fromentin was told by his dealer, Beugniet, that

John Frederick Lewis's 1856 watercolour drawing *A Frank Encampment in the Desert of Mount Sinai, 1842*, 1856, watercolour, shows the visiting Arabs and their camels.

the buying public preferred paintings of horses to those of camels. If one looks at the paintings of Fromentin or the Moroccan paintings of Eugene Delacroix, one finds that horses occupy pride of place in those canvases. Paul Jean Baptiste Lazerges (1845–1902) seems to have been unique in specializing in the portrayal of camels, but little or nothing seems to be known about this obscure Orientalist painter.

Although he was not a specialist in camels, nevertheless camels featured prominently in the oil paintings and water colours of a much greater artist, the British Orientalist painter, John Frederick Lewis (1805–1876). From boyhood onwards Lewis had shown a talent for depicting animals. Though he later turned to interiors and street scenes in Cairo, animals and birds continued to feature prominently in his gem-like watercolours and oil paintings.[14] Four camels appeared in his *A Frank Encampment in the Desert of Mount Sinai, 1842* (1856). When this painting was displayed at the Royal Academy, John Ruskin, that grand arbiter of artistic taste, declared that he had 'no hesitation in ranking it among the most *wonderful* pictures in the world' and he added that the viewer of this picture 'should examine, for instance with a good lens, the eyes of the camels, and he will find there is as much painting, beneath their drooping fringes as would, with most painters be thought enough for the whole head.'[15] But by 1859 Ruskin was getting a little weary of Oriental subject matter and with reference to Lewis's *Waiting for the Ferry Boat – Upper Egypt*, which was being exhibited at the Royal Academy, Ruskin wrote: 'We go to this melancholy Egypt through plague, and mosquitoes, and misery of every sort – and all we see for our pains is a camel with a fine carpet on his back. Cannot we see that any day at the Zoological Gardens?'[16]

SOME LITERARY APPEARANCES

Robert Burton, when he came to write of love-melancholy in his *Anatomy of Melancholy* (1638), picked the camel out as the epitome of sexual jealousy: 'and those old Egyptians, as Pierius informeth us, express in their hieroglyphics the passion of jealousy by a camel; because that, fearing the worst still about matters of venery, he loves solitudes, that he may enjoy his pleasure alone, *et in quocunque obvios insurgit, zelotypiae stimulis agitatus*, he will quarrel and fight with whosoever comes next, man or beast in his jealous fits. I have read as much of crocodiles . . .'. (The Pierius in question was a third-century priest in

Alexandria. Needless to say, the camel was not really a hiero-glyphic emblem for jealousy in ancient Egypt.)[17]

Though the camel was a marvel to Sir Thomas Browne, it was a rather vulgar one: 'ruder heads stand amazed at those pro-digious pieces of Nature, Whales, Elephants, Dromidaries and Camels; these I confess are the Colossus and Majestick pieces of her hand'. But Browne's *Religio Medici* went on to argue that, to the discerning mind, insects were even more impressive ex-amples of the wonders of nature and he further confessed that he found the workings of his own body more interesting to contemplate than the whole of Africa with all its contents.[18]

THE CAMEL IN FICTION

Unsurprisingly, camels feature in one of the greatest novels of the nineteenth century, Gustave Flaubert's *Madame Bovary* (1857). In chapter Six, where the imagery of Emma Bovary's trashily romantic reading is being discussed, we find the following:

> And ye too were there, ye sultans with your long pipes, stretched drowsily in the shade of an arbour in the arms of Bayaderes and giaours, Turkish scimitars. Greek caps, and you, above all, pale landscapes of dithyrambic regions which so often indulge us with a simultaneous display of palms and fir trees, tigers on this side, lions on that, Tartar minarets on the horizon, Roman ruins in the foreground and kneeling camels in the middle distance, the whole within a framework of virgin forest very neatly trimmed, with a great perpendicular ray of sunlight trembling on the water, whereon in patches of white on a steel-gray surface swans are depicted proudly oaring their way far and near.[19]

In the above passage Flaubert was fastidiously describing exactly the kind of novel that *Madame Bovary* was not. He could indeed have written a novel featuring giaours, scimitars, palm trees and camels, for just a few years earlier, from 1849 to 1851, he had toured Greece and Egypt with a friend. He was certainly very familiar with the camels of Cairo and elsewhere and he wrote home as follows: 'One of the finest things is the camel. I never tire of watching this strange beast that lurches like a turkey and sways its neck like a swan. Its cry is something that I wear myself out trying to imitate – I hope to bring it back with me – but it's hard to reproduce – a rattle with a kind of tremendous gargling as an accompaniment.'[20] He was also fascinated with how camel's urine created a glazed paving effect on the sand.

Even before he went out to Egypt, camel-riding was the pivotal image of youthful daydreaming about exotic lands. As he put it in *L'Education sentimentale* (published in 1869, but effectively finished in 1845): 'Oh to feel oneself swaying on the back of a camel!' He identified quite strongly with the beast. A year after his return from Egypt, in a letter to his mistress, Louise Colet, he explained that he cannot change his nature. He is constrained by the gravity of things, 'which makes the polar bear inhabit the icy regions and the camel walk upon the sand'. 'Why the camel?' asked Julian Barnes, who commented on this passage in his novel *Flaubert's Parrot* (1984). Barnes suggests that it is perhaps 'because it is a fine example of the Flaubertian grotesque: it cannot help being serious and comic at the same time'. Barnes also drew attention to the parallel Flaubert made between himself and the dromedary, viz. that he is hard to get going but once is involved in some activity, he is hard to stop.[21]

'"Take my camel, dear," said my aunt Dot, as she climbed down from this animal on her return from high Mass', is how Rose Macaulay's fine and subtle novel *The Towers of Trebizond* (1956) begins. It is one of the most famous opening lines in British fiction. Macaulay continues:

> The camel, a white Arabian Dhalur (single hump) from the famous herd of the Ruola tribe, had been a parting present, its saddle bags stuffed with low-carat gold and flashy orient gems, from a rich desert tycoon who owned a Levantine hotel near Palmyra. I always thought it to my aunt's credit that, in view of the camel's provenance, she had not named it Zenobia, Longinus, or Aurelian, as lesser women would have done; she had, instead, always called it, in a distant voice, my camel, or the camel. I did not care for the camel, nor the camel for me.

According to Laurie, Aunt Dot 'looked back at the open spaces of her youth, when one rode one's camel about deserts frequented only by Arabs, camels, flocks of sheep and Gertrude Bell.'[22]

It is clear that Rose Macaulay knew quite a lot about camels, their diet and load-bearing capacity. I strongly suspect that Macaulay had dipped into Alois Musil's *The Manners and Customs of the Rwala Bedouins*, published in 1928, and that her Dhalur is a misrendering of Dhalul, the term that the Rwala use for a riding camel regardless of sex; as we shall see, 'regardless of sex' is most important for Macaulay's purposes. Musil also noted that pure white camels are the most highly valued among the Rwala. In her seventies Macaulay had travelled in the Levant and there are certain elements of autobiography in the novel. Moreover, Aunt Dot may be based on Macaulay's keen Anglican friend, the crime writer Dorothy Sayers.

When *The Towers of Trebizond* was published, one of its reviews was headlined: 'Mad camel plays a big part in unusual book'. This is true, but the novel is primarily about the loss of Christian faith, loss of the loved one and sexual betrayal. It is a darkly comic story of a Christian mission to Turkey, of Aunt Dot, Laurie and the Revd, the Hon Fr Hugh Chantry-Pigg. Trebizond is both an actual port on the Black Sea and in the end a hazy, glittering symbol of God's mercy, from which Laurie is excluded because of an adulterous affair. The novel followed on immediately from Macaulay's non-fiction book *The Pleasure of Ruins*, and ruins, emblems of change and loss, feature prominently in the novel. But there is an important subplot about the narrator's developing relationship with the camel. For a while, Laurie warms to the beast and finds it fun to ride in front of the hump. And Laurie even thinks about buying a white racing camel to own. Once arrived in Syria and Jordan, 'I and the camel were part of the gorgeous pageant of the East'. Some kind of love affair with the camel, akin to that of the novelist T. H. White with his goshawk or Gavin Maxwell with his otters, seems to be promised. But then the affair seems to cool as the camel shows signs of erotic distraction and madness.

As with the vexed topic of how many children Lady Macbeth had, the sex of Aunt Dot's camel is one of the most puzzling questions in English literature. Its sex is perfectly unclear. The possibility of the camel being called Zenobia suggests female. On the other hand, Longinus as a possible name suggests that the camel is male, but throughout the novel the camel is consistently referred to as 'it'. This difficulty in sexing Aunt Dot's camel functions as a special kind of prolepsis that may lead alert readers to reflect on the difficulty of sexing the narrator and protagonist, Laurie. In fact, Laurie's sex is unassigned until almost the end of the novel. Perhaps the point is that gender identity is not as important as most novelists make it out to be. Or perhaps Macaulay wished to demonstrate that the same possibilities of adventure may be open to women as to men.

Kicked to Death by a Camel by Clarence J.-L. Jackson (New York, 1973) is a whodunnit set in Tamannrasset in southern Algeria, where the protagonist Roger Allenby is researching a book about the history of camels. (As it happens, Clarence Jackson is the pseudonym of Professor Richard Bulliet, who wrote that fine historical study, *The Camel and the Wheel*.) In the novel Allenby says of himself 'I'm a historian. I'm only interested in the role camels played in economic history. I don't even *like* camels very much.' (Perhaps it is also the author who speaks here?) Before Allenby can get very far with his research, Gino Banchero, an Italian tourist, is found dead in the desert. At first sight, it seems that he might have been kicked to death by a camel. But one of the premises of the novel is that the camel's foot is too soft for its kick to kill. The town is full of all sorts of suspect eccentric foreigners and misfits (a bit like the film *Casablanca*), among whom it seems that there is another camel expert Claude Monod, a student of veterinary medicine, who is writing a dissertation on camel diseases. Another suspect, Helmut, has worked on the sedenterization of nomads in Niger.

The novel is enjoyable, but at the risk of offending the learned and brilliant Bulliet, I believe it to be based on a false premise. There are plenty of real-life instances of men being kicked to death by camels, or at least sustaining very serious injury. For example, John F. Keane, a nineteenth-century traveller in Arabia, provided a feeling description of what it is like to be kicked by a camel:

All at once the brute stopped dead, lifted up its nearest hind-leg, as if it were going to scratch the top of its back, and then lashed out, hitting me on the lower part of my chest. I was lifted off the ground and came down on my face on the hard stony ground. The agony was awful; I felt as if my whole inside was torn up; I turned on my back, closed my eyes, and asked people not to touch me . . . I have seen a heavy man fired several yards into a dense crowd by the kick of a camel, and picked up insensible.[23]

Robyn Davidson in *Tracks* also suggests that the kick can be pretty lethal: 'Now a camel can kick you in any direction, within a radius of six feet. They can strike with their front legs, and kick forward, sideways or backwards with the back. One of those kicks could snap you in half like a dry twig.'[24] And is it not the case that the well-known Turkish proverb runs 'The camel's kick is soft, but it takes away life'?

Turning now to something rather different, according to an ancient Kazakh legend, men could be made zombie slaves known as *mankurt*s. The process involved shaving the head of the prisoner before fitting him with a cap of raw camel hide. Then the captive would be left tied to a stake in the desert for several days. As the camel hide dried, it put horrible pressure on the man's skull. Moreover, the hairs of the scalp, unable to grow outwards, grew into the brain, causing great anguish. The man's sanity and memories were squeezed. If he survived this ordeal, he became the slave of his captive.[25] There is a *mankurt* (the son of Nayman-ana) in a Kirghiz science fiction novel, *The Day Lasts More Than a Hundred Years* by Chingiz Aitamov, in which it also features as a metaphor for the plight of those who have forgotten their homeland and culture.[26]

Before becoming a writer, Aitmatov studied at the Animal Husbandry Division of the Kirgiz Agricultural Institute in Frunze and animals feature as the protagonists in many of his stories. *The Day Lasts More Than a Hundred Years* (1980) is a near-future thriller about the encounter of a spaceship from earth with aliens, and it is mainly set in a small Kirgiz town near a Soviet cosmodrome. The story of interplanetary contact is inter-woven with the story of the raising of a pedigree Bactrian called Burannyi Karanar by his Kirgiz master, Burannyi Yedigei. In the course of the book one learns a great deal about the hard-ships of rearing camels on the Kirgiz steppe and the drinking of *shubat* (alcoholic fermented camel's milk).[27] Karanar fights the older head of the herd to take over and rut. [28] Once the rutting season has begun, Yedigei finds it impossible to control his lust-inflamed camel. In mid-winter Karanar escapes, and gallops like a fury let loose on the steppe. He accumulates four she-camels and he defends his harem by pissing with fury. Yedigei has to chase after him and fights him with a whip. He succeeds in subduing his camel, but when he returns, he dis-covers that the woman that he desired has vanished while he was on his camel hunt. So he takes it out on Karanar with a whip and then chases him away. The camel returns worn out, skin and bones.[29] As in *The Towers of Trebizond*, one's sympa-thies are with the camel.

KIPLING AND OTHERS

Rudyard Kipling (1865–1936) published a five-verse poem, 'Oonts', in 1890. *Oont* is Hindustani for camel. In the poem Tommy Atkins, stationed in India, laments the troubles he has had with camels. Here is a sample from the lament:

O the oont, O the oont, O the commissariat oont!
With 'is silly neck a bobbin' like a basket full o' snakes;
We packs 'im like an idol, an' you ought to 'ear 'im grunt,
An' when we gets 'im loaded up 'is blessed girth-rope breaks.

In this blackly comic, demotic account of the beast, the commissariat camel is a devil, an ostrich and an orphan all in one. His smell is vile. He is more dangerous than a Pathan tribesman, and so on.[30]

Camels also provided the background furnishings to Kipling's famous poem, 'The Ballad of the King's Jest'. The opening verses are as follows:

When spring-time flushes the desert grass,
Our kafilas wind through the Khyber Pass.
Lean are the camels but fat the frails,
Light are the purses but heavy the bales,
As the snowbound trade of the North comes down
To the market-square of Peshawur town.

In a turquoise twilight, crisp and chill,
A kafila camped at the foot of the hill.
Then blue-smoke haze of the cooking rose,
And tent-peg answered to hammer-nose;
And the picketed ponies, shag and wild,
Strained at their ropes as the feed was piled;
And the bubbling camels beside the load
Sprawled for a furlong adown the road;
And the Persian pussy-cats bought for sale,
Spat at the dogs from the camel-bale;
And the tribesmen bellowed to hasten the food;
And the camp-fires twinkled by Fort Jumrood;

'The Djinn beginning the Magic that brought the humph to the Camel' in Kipling's story 'How the Camel Got His Hump' from *The Just So Stories*.

And there fled on the wings of the gathering dusk
A savour of camels and carpets and musk,
A murmur of voices, a reek of smoke,
To tell us the trade of the Khyber woke.[31]

(A kafila is a caravan train.)

But Kipling is better known for his short tale, 'How the Camel Got His Hump' which appeared in *The Just So Stories* (1902). To summarize it as briefly as possible, in the early days of the world a Camel lived by itself in the Howling Desert in complete idleness and when any other creature spoke to him, he just responded with 'Humph!' The Horse, the Ox and the Dog, all of whom were already working for Man, became angry at the Camel's getting away with doing nothing. They took their complaint to the Djinn. The Djinn went into the Desert to remonstrate with the Camel, but the only response he got was 'Humph!' So he set to work to create great magic. Then when he next asked the Camel to go to work and got the response of 'Humph!' the Camel 'saw his back, that he was so proud of, puffing up and puffing up into a great lolloping humph'. The Djinn then tells the him that he must now go to work and that the humph (these days spelt 'hump' so as not to offend the camel) will allow him to work for three days without eating. The story is followed by a set of moralizing verses, of which the first runs as follows:

> The Camel's hump is an ugly lump
> Which you may well see at the Zoo;
> But uglier yet is the hump we get
> From having too little to do.[32]

There is something about the camel that makes it the victim of light verse. This is Hilaire Belloc:

> The dromedary is a cheerful bird.
> I cannot say the same about the Kurd.

NON FICTION

Charles Doughty's strange classic of travel writing, *Arabia Deserta*, which described his difficult travels in north-west Arabia and the Najd in the years 1876–8, was first published in 1888. Doughty was passionate about geology, which he regarded as proof of God's handiwork in the slow shaping of the world. He also wished to shame the benighted heathen of the Arabian Peninsula by displaying to them the comportment of a Christian gentleman. Above all, he undertook his adventure in order to provide material for an English epic that would revive the lost glories of English prose. However, the camels kept obtruding. Here he describes how baggage camels are unloaded by the tribesmen he was travelling with:

> The bearing camels they make to kneel under their burdens with the guttural voice *ikh-kh-kh* ! The stiff neck of any reluctant brute is gently stricken down with the driving stick or an hand is imposed upon his heavy halse; any yet resisting is plucked by the beard; then without more he will fall groaning to his knees. Their loads discharged, and the pack-saddles lifted, with a spurn of the master's foot the bearing-camels rise heavily again and are dismissed to pasture.[33]

A curious story of a phantom camel was related by Doughty in his characteristically contorted and arcane prose:

> These few hill-men, not forsaking the old hospitality, are, we have seen, commended by the tribes: yet there was a strange tale told, at this time in their tents. 'A certain Belùwy or Bíllî tribesman, was going over the Harra; at the sun setting, where he halted to pass the night, a strange camel appeared to him, standing over him, and the camel uttered a manner of human speech, 'These murrains and the great drought they come oftener upon you, and the locusts, not as beforetime, but now year by year, and ye

wot not wherefore:— wherefore go the Beduw back from the custom of the fathers? Ye suffer the wayfaring man to pass by your byût, and the hungry man goes from you empty!' The Arabs spoke of the phantom by twos and threes in their tents and in their mejlis, and this was now a tale current in all the country. Some asked me – a book man – 'how I looked upon it?' all the people who knew him who had seen the phasm, to be 'a good understanding man.'[34]

Doughty does not relate what his reply was.

Doughty rarely got on with the Bedouin and camels were often an issue here. 'I bought thus upon their trust, a dizzy camel, old and nearly past labour and, having lost her front teeth, that was of no more value, in the sight of the nomads, than my wounded camel. I was new in their skill; the camels are known and valued after their teeth, and with regard to the hump.' For those who know and care about camels, *Arabia Deserta* is a sad book, for Doughty was again and again conned into buying clapped-out, toothless and arthritic camels and he was ruthlessly mocked by the Bedouins for his naivety.

T. E. Lawrence was sharper and he made a point of riding only the best camels. *The Seven Pillars of Wisdom* (1926) has been read as a history of a First World War campaign, a ripping yarn, an individual psychodrama, or a piece of deceitful imperialist apologetics. But another valid way of reading Lawrence's book is as an informally structured encyclopedia of the camel written in prose poetry. He describes their vital role in the Bedouin economy. He evokes the ritual of the drum beats to signal the loading of camels and the second drum roll to signal that the caravan is moving off. He comments on the gregarious nature of the camel, and gives guidance on how to mount a camel, and on how to make it kneel by tapping it on the neck. He gives instructions to assuage a mother camel's grief by presenting her with the hide of her calf, and on how to alleviate mange with butter and how to cook a suckling calf. He describes their favourite forage and the difficulties camels have with night journeys and with snowy ground. Lawrence was a precise observer and his description of how camels kneel has already been quoted.

It is not clear whether he had ever ridden a camel before he entered the Hejaz. (Quite likely he had when he made his undergraduate trip round crusader castles, or later when he worked on the excavation at Carcemish.) It was a love affair which had its ups and downs. In March 1917, early on in his time with the Bedouin, Lawrence, who was feverish and exhausted from the day's travelling, lay down to rest. The camels were first set free to graze on stubble grass:

A Bactrian camel
tethered on its
side.

The camels loved this grass which grew about sixteen inches high, in tufts, on slate green stalks: and they gulped down great quantities of it until the men drove them in and couched them by me. At the moment I hated the camels, for much food made their breath stinking, and they belched up a new mouthful from their stomachs each time they had chewed and swallowed the last, till the green slaver of it ran out between their loose lips over the side-teeth and dripped down their sagging chins.

In my anger from where I lay I threw a stone at the nearest which got up and wavered about my head near me for a time: finally it straddled its back legs and staled in wide bitter jets and I was so far gone with the heat and weakness and pain that I just lay there and cried about it unhelping.[35]

He was to become a connoisseur of camels. One of his companions possessed an enviable beast: 'Ghazala, a camel built like an antique, grand and huge, towering a good foot above the next in size of all our animals; and yet perfectly proportioned and with the stride like that of an ostrich: a lyrical beast, the noblest and best-bred of the Howeitat camels, a female of nine remembered dams.'[36] Lawrence was filmed and photographed on or beside camels so much that one tends to forget that a lot of the time he was driven about the desert in an armour-plated Rolls Royce, as S. C. Rolls revealed in his memoir *Steel Chariots in the Desert*.[37]

CELLULOID CAMEL

Turning now to films, the camel featured in many early ethnographic documentaries. But possibly its first appearance in a feature film was in 1924 when in René Clair's experimental

Entr'acte (a film designed as an interlude in a Dadaist ballet), a dromedary was employed to pull a hearse through the streets of Paris. Eventually the dromedary succeeds in biting through the rope that attaches it to the hearse and the hearse hurtles down a hill with the mourners in headlong pursuit. The appearance of the camel in Clair's short film set a pattern for the future, as the camel has mostly featured in films as a comic accessory. For example, *The Road to Morocco* (1942) featured Bob Hope, Bing Crosby, Dorothy Lamour and two talking camels. (The female is called Mabel.) Let a talking camel have the first word about this film: 'This is the screwiest picture I was ever in.' Cast ashore on a desert coast, Bob and Bing encounter a talking Bactrian (*sic*). At one unscripted point, the ad-libbing Bactrian kisses Bob on the back and then spits in his face.

Hawmps! (1976) was a comedy featuring the US cavalry on camels, which billed itself as 'Very possibly the funniest motion picture of the decade', though *Halliwell's Film and Video Guide* has characterized it as an 'incredibly overstretched and tedious period comedy'. The cavalry outpost in the Wild West had been promised a delivery of fresh 'Arabians'. They were expecting horses, but what they got was camels. Quite a few of the camel jokes in the film are about the cigarette brand, rather than the animals. Then there was *Ishtar* (1987), starring Dustin Hoffman and Warren Beatty, a (would-be) comedy film about two untalented song writers in the Saharan republic of Ishtar. This has featured on many lists of the ten worst films of all time – and was also one of the most expensive. In its review of the film *Daily Variety* commented: 'One can't help but wonder whether the camel was the only blind creature who had something to do with this picture.' The talentless duo end up on the run in the desert with the aforementioned blind camel, which Lyle (Beatty) had purchased in a camel market as some kind of espionage password. This film lost a lot of money. Apparently large sums were expended in looking for a blue-eyed camel that would look blind on film.

The camel has featured in a few more serious films. Most of David Lean's *Lawrence of Arabia* (1962) was shot on location in the Jordanian desert and local Bedouin were recruited as extras. The actors headed by Peter O'Toole were given lessons in camel riding. O'Toole, having got a sore bottom from so much riding, when he had a few days off in civilization, purchased a layer of foam which thereafter he put under the saddle, whereupon the Bedouin called him Abu Isfanj (The Father of the Sponge); several of them took to following his example and this can be seen in the film. Peter O'Toole formed a bond with his No. 1 camel, Shagram, and on one occasion Shagram saved his life.

T. E. Lawrence and a distant camel in David Lean's 1962 film *Lawrence of Arabia*.

Desert Patrol Camels – of which Shagram was one – are trained to stand over their riders if they are injured, and protect them. When one of the final scenes in Jordan was being shot, Shagram stumbled over some rocks in charging downhill, and Peter was thrown off, striking his head.

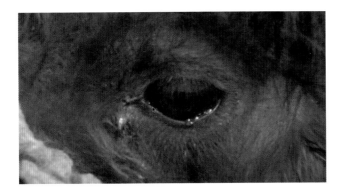

Stills from Byambasuren Davaa and Luigi Falorni's 2003 film *The Story of the Weeping Camel*.

He fell in the path of five hundred Arab horsemen, coming straight towards him. Although the Arab riders were all drawn from cavalry regiments, they were so massed they could hardly have avoided Peter. Shagram, however, stood her ground and the actor avoided further hurt as the Arab horses thundered past.[38]

The camels were predictably unpredictable. In one scene a Bedouin called Ali was remonstrating with Lawrence about the danger of crossing the waterless desert to Aqaba and Ali's camel was expressing his firm agreement with a series of snorts, which are a mixture of a hoot, a particularly nasty example of whooping cough and a bellow of pain.

> 'If we go on,' declared Ali, 'the men will begin to die.'
> Ali's camel let out a specially horrified and mournful explosion.
> 'And if we still go on?' enquired Lawrence mildly.
> 'The camels will begin to die,' said Ali firmly.
> Whereupon his camel gave out the loudest and most blood-curdling noise ever heard from such a beast.[39]

Subsequently additional scenes were shot in Morocco, but this caused problems, as Moroccans rode camels in a different way from Jordanians and rode on a different kind of saddle. Hundreds of Jordanian-style saddles had to be swiftly manufactured and the Moroccans were taught to ride like Jordanians. (Moroccans normally sit in front of the hump, on the camel's shoulder.)

The Story of the Weeping Camel (2004), a moving ethnographic documentary filmed in Mongolia by Byambasuren Davaa and Luigi Falorni, is the story of how a mother, who, after a difficult delivery has rejected her white colt, is with great difficulty induced to recognize it and allow it to suckle. The Mongolian herdsmen use violin music to heal the breach between mother and colt. An expert violinist is fetched from a distant village, who performs an ancient bonding ritual and the strains of his violin persuade the mother to weep and accept her colt. The faces of the camels are simultaneously noble and stupid. Their groans are melancholy. The three-generation family at the centre of the film tend 300 sheep and goats and 60 camels. The hardness of life as a Mongolian pastoralist on the windswept Gobi desert with no company but dust devils is powerfully evoked. The film was touch-and-go as the makers only had a limited stock of film and, as Falorni observed, 'we found it impossible to direct a camel'.

Finally, those awful noises that Chewbacca makes in the *Star Wars* films are alleged to have been modelled on the groaning of a camel.

6 The Camel's Role in History

'A man could powerfully imagine himself the lord of creation
when mounted on a camel.'
Geoffrey Moorhouse

BIBLICAL BEAST

It is impossible with any confidence to date or give details for
the history of the domestication of the camel. The evidence is
not there. But by comparison with sheep or dogs, it would seem
that camels were domesticated relatively recently. Although
there is early evidence of ancient peoples making use of camel
hair, dung and bones, this does not prove that the animals that
provided these materials were domesticated. The early inhabitants of Palestine relied on asses rather than camels.[1] Though
there is clear evidence for its domestication in Palestine by the
eleventh century BC, apparent earlier references to the camel in
Genesis and Exodus are probably anachronistic. But the debate
about how early the camel was domesticated in the Middle East
is a contentious one, as supporters of the literal truth of Bible
struggle (with some suggestive evidence) to produce evidence
that Abraham could have owned camels. (According to Genesis
12:6, Pharaoh offered Abraham camels.)

The camel is mentioned 62 times in the Bible and is the tenth
most frequently referred-to animal (56 of those references are
found in the Old Testament). When the Queen of Sheba came to
visit Solomon, her baggage was carried on camels (Kings I, 10:2.)
In Jeremiah 2:23 Israel is described as whoring after false gods

like a camel on heat. In the New Testament, we find: 'It is easier for a camel to pass through the eye of a needle than for a rich man to enter the kingdom of heaven' (Matthew 19:24). Although it has been suggested that this striking simile arises from a misreading of the Hebrew for a rough thread, which is similar to the word for camel, it seems more likely that the camel reading is correct, as when the disciples heard Jesus's words, 'they were exceedingly amazed'.

SPREAD OF DOMESTICATION

There were wild camels in North Africa, but these were almost entirely exterminated in the third millennium BC. Camels were not used in ancient Egypt or in Carthage. It is hard to find conclusive evidence, but Trevor Wilson, an animal scientist specializing in the arid zones, believes that the camel was first domesticated in south Arabia 4,000 years ago.[2] It should be borne in mind that there is domestication and domestication.

An Assyrian stone relief of woman with camel from Nimrud, Mesopotamia, c. 850 BC.

At a simple level, camels may have first been herded for their meat and milk and only later may they have been ridden and used for the transport of spices and other things in Arabia.[3]

The Bactrian camel was probably domesticated around the first quarter of the first millennium BC in Turkestan. The Bactrian did not replace wheeled traffic as a means of transporting goods but coexisted with it. It also seems that it was not used as a pack or riding animal, but was rather used to pull carts. The Uighurs used Bactrians as farm animals. In Tang China there was much demand for camels as beasts of burden and also as suppliers of meat and hair. The eventual use of the Bactrian as a pack animal made the Silk Road from China to Rome possible. The route ran for almost 5,000 miles roughly from Loyang to Samarkand to Tashkent to Antioch. The route opened up in the second century BC, but it really became important from the first century AD onwards. At first only Bactrians were used, but later cold-adapted dromedaries were also employed. Although the Bactrians have been used for transport and other forms of labour, the tribes

A Chinese model camel found on the 'Silk Road'.

who herd them tend not to depend upon them for subsistence, in the way that some Arabs and East Africans relied on dromedaries. In medieval times Bactrians were common in Iran, Iraq, Anatolia, India and Central Asia, but today they are more or less restricted to north-east Afghanistan, China and Mongolia.[4]

Pastoral camel nomads first become an important factor in Middle Eastern history around 1000 BC.[5] People began to use dromedaries for warfare. The Achmaenid king Cyrus the Great (559–530) relied heavily on camels to bring up supplies. They were also used by him in battle. According to Herodotus (c. 484–425 BC), when Cyrus confronted Croesus in eastern Turkey, he took the baggage off his camels and mounted his troops on them and sent them against the enemy cavalry because he knew the horses would be afraid of them. Elsewhere, in his *Histories*, when describing desert-dwelling Indians, Herodotus tells the Greeks what they apparently already knew about camels:

> The sand has a rich content of gold, and it is this that the Indians are after when they make their expeditions into the desert. Each man harnesses three camels abreast, a female on which he rides, in the middle, and a male on each side in a leading rein, and takes care that the female is one who has recently dropped her young. Their camels are as fast as horses, and much more powerful carriers. There is no need for me to describe the camel, for Greeks are familiar with what it looks like; one thing, however, I will mention, which will be news to them: the camel in its hind legs has four thighs and four knees, and its genitals point backwards to its tail.[6]

The ambitious Roman consul and general Crassus was defeated at Carrhae in eastern Turkey in 53 BC because the Parthians

A man leading a
camel, or perhaps
a giraffe, from a
5th-century AD
Roman mosaic
floor.

(Persians) had camels and he did not. Parthian archers on camels
and horses rode around the flanks of Crassus's army. Camels
also played a crucial role in bringing up more arrows for the
rapidly firing mounted archers. (Crassus had been hoping that
the archers would soon run out of arrows.) The Emperor Trajan
(98–117), who spent a lot of time fighting the Parthians in
Persia, founded a camel corps, Ala I Ulpia Dromedariorum
Milliaria, which was based in Palmyra.[7] On a more sporadic basis
camels were quite widely used by the Romans and, for example,
camel bones have been found on the site of a Roman villa in
Soissons in northern France.[8]

The use of domesticated camels spread to North Arabia,
Palestine and Syria. The wild camel in North Africa had become
extinct before 3000 BC. The expansion of the Sahara Desert
would have increased the demand for camels and it is probable
that the camel was reintroduced into North Africa in Roman
times. The first written reference to camels in North Africa occurs
in 46 BC when Caesar defeated Juba, a North African ally of

Pompey, and captured 22 dromedaries. After the fall of Rome and the deterioration of settled agriculture in North Africa, many Berbers switched to nomadism and the population of camels presumably increased. It seems likely that the raids carried out by nomads on settled agriculturalists contributed to the desertification of North Africa. The African emperor Septimus Severus (r. 193–211) encouraged camel breeding in North Africa and from then onwards camels became common in the region.[9]

CARAVAN CITIES

From the second century BC the Arabs took over the caravan trade of Syria and Iraq and much of this trade was in profitable spices. This was part of the background to the rise of Nabataean Petra in what is now Jordan. Nabataeans, who may or may not have been of Arab descent, appear in history in the fourth century BC. They controlled a series of oases in southern Jordan and north-west Arabia and they became masters of the camel-borne caravan trade. The Greek geographer Strabo called them

Mid-2nd century AD mosaic of Silenus on a camel, from El Jem, Tunisia.

'hucksters and merchants'. From the first century BC onwards the caravan city of Petra became their capital. In a rare external literary reference to the inhabitants of Petra, Diodorus Siculus in the first century BC wrote that 'some of them raise camels, others sheep, pasturing them in the desert'. According to Strabo, they had huge caravans that traded in frankincense and myrrh. This they purchased in south Arabia, but since transport by ship was cheaper than by camel, the spices were then ship-borne until the Gulf of Aqaba, where winds and tides are unfavourable. There the spices were offloaded on to camels, which carried them on to Alexandria, Gaza and Damascus.[10] The Nabataeans adopted the North Arabian saddle for their camels. The first representation of this saddle is on a Roman coin of *c.* 50 BC showing the surrender of a Nabataean king.[11] In 106 AD the Nabataeans were conquered by Trajan and Petra was incorporated into the Roman empire.

By about 100 BC Palmyra in the Syrian desert had become an important stopping point for camel caravans passing between Persia and the Mediterranean ports, but it only became really wealthy in the following century. As a city state, situated between the Roman and Parthian empires, it prospered from the camel caravans. Arsu, the Palmyran god of the evening star and of caravans which were guided at night by that star, is usually portrayed riding a camel with his twin brother Azizos, the god of the morning star. The Palmyran camel corps also used the North Arabian saddle. Palmyra favoured the camel and imposed punitive taxes on cart loads entering the city – four times the amounts paid on camel loads. Palmyra became part of a Roman province under Tiberias (AD 14–37), but it continued to flourish as a commercial city until about AD 270.[12]

It is thought that when camel herders first began to ride upon their camels, they rode upon the South Arabian type of camel saddle. Thesiger gave a characteristically meticulous description of the South Arabian saddle:

> The southern Bedu ride on the small Omani saddle instead of the double-poled saddle of northern Arabia to which I was accustomed. Sultan picked up my saddle, which was shaped like a small wooden vice, fitted over palm-fibre pads, and girthed it tightly over Umbrausha's withers just in front of the hump. This wooden vice was really the tree on which he now built the saddle. He next took a crescent-shaped fibre pad which rose in a peak at the back and, after fitting it round the back and sides of the camel's hump, attached it with a loop of string to this tree. He then put a blanket over the pad, and folded my rug over this, placed my saddle-bags over the rug, and finally put a black sheep-skin on top of the saddle-bags. He had already looped a woollen cord under the camel's stomach so that it passed over the rear pad, and he now took one end of this cord past the tree and back along the other side of the saddle to the original loop. When he drew the cord tight it held everything firmly place in place. He had now built a platform over the camel's hump and the fibre pad which was behind it. Sitting on this the rider was much farther back on the camel than he would have been if riding on the northern saddle, which is set over the camel's withers.[13]

With the South Arabian saddle (in Arabic *rahl*) the rider sits behind the hump and guides and drives the camel with a stick.

A double-horned North Arabian saddle (*rahl*).

The North Arabian saddle (in Arabic *shadad*) was introduced around 500 to 100 BC and only then was the camel as a pack animal able to compete with wheeled transport. The North Arabian saddle consists of two inverted wooden Vs placed on either side of the hump and connected to one another by four crosspieces and surmounted by a pad. This saddle surrounds the hump and sits on the ribcage. Though the rider sits above the hump, his weight is borne by the camel's ribcage. When loads are placed on a camel, they are distributed on the sides of the North Arabian saddle. The camel's load-bearing capacity was thereby increased. The North Arabian saddle made it easier to fight with a sword or lance from camelback and gave the warrior the advantage of increased height.

Richard Bulliet's *The Camel and the Wheel* (1976) is a groundbreaking historical study of how the camel replaced the cart and chariot in the Middle East and North Africa and how the invention of the North Arabian saddle gave camel-rearing nomads a

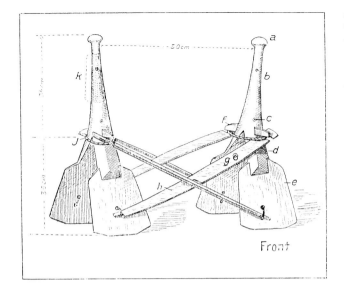

A North Arabian camel saddle, from Alois Musil, *The Manners and Customs of the Rwala Bedouins . . .* (1928).

crucial edge in commerce and warfare. Since the North Arabian saddle increased the camel's load-bearing capacity, camels as pack animals began to compete with and displace wheeled traffic in the Middle East and North Africa. Also, the camel can carry twice the load of a donkey. Previously the Romans in the East had relied on wheels for the transport of goods and, when they used camels, they used them for pulling carts or ploughs. The switch from wheel-borne traffic to the use of pack animals seems to have occurred sometime between the third and seventh centuries AD. The decline of the Roman road system was both a cause and a consequence of this switch. With the change to pack animals there was less need for wide streets in the cities and so one saw the rise of cities, like Cairo or Aleppo, with twisting maze-like layouts of streets. Roads ceased to be paved because that was more comfortable for animals. Camels were much happier treading on earth or sand. The carts and chariots that were common in the Middle East in Roman times disappeared. The near total absence of wheeled traffic in the pre-modern Arab world had long been observed.[14] According to the eighteenth-century philosopher and traveller Volney, 'It is noteworthy that in the whole of Syria no cart or wagon is to be seen.'

The importance of the nomads on the margins of the Byzantine and Sassanian empires increased, as they were the chief suppliers of pack animals for long distance trade and military transport. Some Arabs went into the caravan trade themselves. Grazing camels were easy to steal and hard to defend, and this fostered a culture of intertribal raiding in the Arabian Peninsula. The most aristocratic tribes were those nomads who herded camels deep in the Arabian deserts, remote from any external political control by the Byzantines, the Sassanians or their client states. The rise of Islam can be seen as a manifestation of the growing politico-military importance of nomads mounted upon camels. It used to be believed that Mecca, which was to be the birthplace of Islam, had previously grown rich from being a crucial stopping-place on the northward route of camel caravans carrying spices from Yemen towards the Mediterranean. However, it has been shown by Patricia Crone in *Meccan Trade and the Rise of Islam* (1987) that Mecca was not really on such a route and there is little or no evidence for any land-borne spice trade in Arabia in the centuries immediately prior to the rise of Islam. Moreover, Bulliet has calculated that a camel caravan progressing up the length of the western side of the Arabian Peninsula would stop about 65 times before reaching the Mediterranean. Even if Mecca had been on the coast route, which it was not, there was no reason for it to have become a privileged stopping-point. Anyway, it would actually be more convenient for a caravan to go by Taif. Mecca was about a hundred miles out of the way.

Patricia Crone also showed that there was really no evidence for a flourishing spice trade in the seventh-century Hejaz. Instead later traditions suggested that there was a local trade in camels, donkeys, leather and camel-hair products. More recently she has published an article arguing for the military significance of livestock rearing in the region. In the sixth and seventh centuries the Roman army was expanding its operations in Greater Syria and Iraq, where it was campaigning against the Sassanid Persians. It was a huge consumer of leather goods: 'tents, scabbards, shields, shield covers, baggage covers, kit bags, purses, horse armour, saddles, reins and other horse-gear, sandals, boots, belts, wine skins, water skins, as well as diverse slings, strings, laces and straps for use in arms and clothing'.[15] As another historian of the Middle East, Lawrence Conrad, has written, 'leather was the plastic of the age'. The Quraysh, who were the dominant tribe in Mecca, and other Arab tribesmen grew prosperous by providing the Roman army with leather, much of that leather coming from camel hides. The intensification of fighting between the Romans and the Persians was an economic opportunity for the Arabs. They also became familiar with the strengths and weaknesses of the Roman army.

ISLAMIC CONQUESTS

The camel was used by Arab armies for carrying baggage and riding towards battles, but they generally fought on foot or on horseback. (There is even an Arabic verb for the practice of dismounting from a camel and then mounting a horse: *tanazala*.) It is very difficult to direct a camel in a sustained gallop and a camel charge lacks the impetus of one delivered by cavalry mounted on horses. So camels played a subordinate, though still important role in the Islamic conquests.[16] The Arab conquest of the Middle East, North Africa and part of Central Asia in the seventh and eighth centuries encouraged the increased use of camels in all those regions and there are correspondingly fewer references to donkeys. Though the Arabs tried several times to conquer Constantinople, they failed and they also failed to occupy Anatolia permanently. In part the latter was surely because the Arab armies were relying on the dromedary, which, unlike the Bactrian, was poorly adapted to altitudes and cold climates. In the Abbasid period, from the late ninth century onwards, there are fewer references to mules and more to camels. The state postal network was using swift she-camels called *jamazat*. A network of camel stations was established under the Abbasids and continued under the Buyid and Samanid successor regimes. Camels required fewer relay posts, since they have more endurance, whereas mules would need to be changed quite frequently.[17]

An illustration in a 12th-century Byzantine history showing an Arab encampment.

The earliest Muslim cavalry forces, whether mounted on horses or on camels, probably did not know of the stirrup, but perhaps by as early as the end of the seventh century the stirrup was beginning to be adopted. Among other things, it gave mounted archers more support.

From the eleventh century onwards the Turkish tribes moved westwards and began to occupy of the Iranian and Anatolian plateaux. These Turks tended to use Bactrians rather than dromedaries and hence they were more successful than the Arabs had been in settling in Anatolia. Characteristically the camel-rearing Turkish nomads were transhumant, moving with their herds from summer grazing in the highlands to winter quarters lower down. They were known as Yürüks. (The word literally means 'walkers'.) From the sixteenth century onwards the Ottoman Turkish Sultanate found itself frequently in conflict with the nomads, whom it tried to settle or deport to places like Cyprus. The nomads were pushed out into marginal territories of high ground, desert or marshland.

MAMLUK AND OTTOMAN CAMELS

The Mamluk regime presided over Egypt and Syria from the 1250s until the Ottoman conquest of those two regions in 1517. A mamluk was a white slave soldier, usually of Turkish origin. Camels were used for logistical loads by the Mamluk army. Every mamluk taking part in a military campaign received at least one camel, sometimes two. The *halqa*, the non-elite Mamluk force, received three camels per two men. The camels were kept in a camel-stable (*munakh*) attached to the royal stables, as were the stables of the dromedaries (*al-hujun*) and she-camels (*al-niyaq*). In the late years of the Mamluk Sultanate a camel cost seven dinars.[18] Under the Mamluks five load-bearing camels and one

Erhard Schoen,
hand-coloured
print of a Mamluk
on a dromedary,
1530.

lead camel were regularly used to bring snow from Damascus to Cairo to cool the sultan's drinks. Every other day, from the beginning of June till the end of the November, the snow detachment left Damascus. Though they were not as fast as horses, camels were also used by the *barid*, the state's postal and intelligence system. The royal stables provided camels for staging posts in Egypt.[19]

In the early 1430s a Frenchman, the Burgundian Bertrandon de la Brocquière, travelled in Mamluk Syria:

I met near Damascus, a very black moor who had ridden a camel from Cairo in eight days, though it is usually sixteen days' journey. His camel had run away from him; but with the assistance of my moucre [dragoman], we recovered it. These couriers have a singular saddle, on which they sit cross-legged; but the rapidity of the camel is so great that, to prevent any bad effects from the air, they have heads and bodies tightly bandaged.

The courier was carrying orders from the Sultan in Egypt to arrest all the Catalans and Genoese who might be found in Damascus and Syria. Just a day later Bertrandon entered Damascus and was in time to witness the return of the hajj caravan from Mecca. The caravan, which was composed of three thousand camels, took two days and two nights to enter the city.[20] Such great caravans to Mecca, which continued into the twentieth century, included international traders as well as pilgrims. From the sixteenth century onwards the need to protect the hajj caravans from Cairo and Damascus from attacks and extortion by the Bedouin was a major preoccupation of the Ottoman government. Customarily an escort of several hundred troops was provided as well as various sorts of caravan specialists. In order to avoid the heat of the sun, much of the journey was done by torchlight.

Sigoli, an Italian pilgrim who visited Egypt in 1384, reported that camel meat was on sale in Alexandria.[21] Felix Fabri, a pilgrim who visited Egypt in 1483, included in the account of his travels a fairly lengthy description of the camel in which detailed and accurate observation mingled with learned misinformation from literary sources. It opened with the statement that 'Among us the camels are considered monsters . . .'. Fabri distinguished between the Bactrian and the Arab camel, but got them muddled up, as he claimed that the Arab camel was the two-humped one. He went on to provide a string of references to the Bible, *The Life of St Hilarion* and Vincent of Beauvais' *Speculum Naturale*. Fabri considered the camel to be a deformed

A drawing of a camel from Arnold von Harff's 1490s *Pilgrimage . . . through . . . Syria, Egypt, Arabia, Ethiopia, Nubia, Palestine, Turkey . . . .*

animal. The camel had big frightening eyes, but he always had a sad and worried expression. The camel's eyes magnified everything that he saw and that is why he seemed so worried. A man advancing on a camel appears as being four times his actual size. The camel can live to be a hundred. Camels respect family relationships and taboos and Fabri had heard of one camel that was tricked into having sex with his mother who was hooded. When the mother's hood was removed, her son was so shocked at what he had done that he committed suicide by biting himself to death. Camels hate horses and donkeys. Cameleers have no need to use a whip or a stick as they drive their camels along by chanting 'Han nay o an no ho ho oyo o ho, etc'. Fabri had examined the camel's feet and compared them to the webbed feet of a goose. The cameleers employed on his journey had been hired by the dragoman from small villages in Palestine, but these Palestinian villagers were not considered proper Arabs by the nomadic Arabs.[22] Fabri was one of the very few medieval writers to give Western readers a true idea of what a desert was like, for many medieval Christians imagined that the Saracens dwelt amidst luxuriant forests and fields.

The pilgrim Arnold Von Harff, who made a pilgrimage from Cairo to the Monastery of Mount Sinai in the 1490s, included in his travelogue a copy of the contract that he made with his cameleer:

> I. N. Mokarij will carry this Frank (so they call us who come from our countries) from here in Cairo to the monastery lying below Mt Sinai on a good camel, on which he shall sit on one side in a wooden box covered with a thick pelt and carrying on the other side his provisions and the camel's food. I shall carry also for him two udders, namely goat-skins, full of water for him, myself and the camel. In addition I will assist him to get on and off the camel, and will stay by him by day and night and attend to his welfare. This Frank N. is to give me two seraphin, namely two ducats, one at Cairo and the other when we reach the monastery below Mt Sinai.

In practice Von Harff found that additional presents or bribes were necessary.[23]

EUROPE

The historic role of the camel was not restricted to Asia and North Africa. There are a few references to camels in early medieval Europe. The Visigoths and other tribes may have brought

them into western Europe. In France the Merovingian king
Clotaire II (d. 629) paraded his Queen Brunehaut on a camel
before having her executed.[24] The Arabs and Berbers who in-
vaded and occupied Spain and the South of France in the early
eighth century brought camels with them, but camel herding
never really flourished in those regions.[25] The Hohenstaufen
king Frederick II made use of camels in Sicily and southern Italy.

There were several attempts to introduce camels into Europe
in the early modern period. Around 1623 a small herd of camels
owned by King James I used to graze daily in St James's Park.[26]
Philip II of Spain (1527–1590) maintained a small zoo in the
gardens of his palace at Aranjuez with four camels, which he
had brought over in the 1570s from Africa. They proved useful
in building work, so more were bred until there were about
forty.[27] Ferdinand II de Medici, Grand Duke of Tuscany, introduced imported camels in 1622 to be used a pack animals. The
last of the herd lingered in the environs of Pisa until the Second
World War, when soldiers killed the remnant for meat.[28] In
nineteenth-century Spain there were feral camels in the swamps
of the Guadalquivir Delta. Allegedly they had been left there
by the British army in the Peninsular War. Alternatively and

less romantically, they had originally been imported into the province of Cadiz in 1829 to work on road-building and other projects. There were other short-lived attempts to introduce camels in Spain, Poland and elsewhere.

AUSTRALIA

H. M. Barker's *The Camels and the Outback* (1964) is the key text on the introduction of camels into Australia. The author 'thought a book about camels might be useful if the hydrogen bomb burst and this is what he wrote'.[29] The first camel arrived in Australia in 1840. More were imported to Victoria in 1860. In 1865 there was a large-scale importation from Karachi. Camels were used by the explorers Burke and Wills on their ill-fated expedition of 1860–61, which attempted to cross Australia from Melbourne in the south to the Gulf of Carpenteria in the north. For this expedition they had three Indian sepoy camel drivers and 26 camels. Early on they had got rid of the only really expert manager of camels, which may have sealed the doom of the expedition. There had been tales of a lush interior, even an inland

Camels were used on Robert Burke and William Wills's ill-fated 1860–61 attempt to cross Australia from south to north.

sea, sited somewhere within the desert that turned out to be the purest fantasy. Although Burke and Wills did get to within a few miles of the Gulf of Carpenteria, the return journey was mismanaged and, when the last camel died, there was no hope for the men. Burke, Wills and five others died.

Later camels and cameleers played a crucial role in the exploration and opening up of Australia. Camels were used as haulage animals for the construction of the Overland Telegraph Line, which ran across the continent from Adelaide to Darwin, and the Transcontinental Railway. Alice Springs was originally a camel junction. The so called 'Afghans' or 'Ghans', imported to look after the camels, were not Afghans, for they mostly came from Baluchistan and Karachi. But 'Ghan town' was the term for the cameleers' quarter. At first, the Aborigines were terrified by the strange animals, but later many of them also became camel handlers. There were perhaps as many as 20,000 domesticated dromedaries in the early 1920s. But the coming of cars and trucks made the herds worthless and the Ghans more or less unemployable. Camels were released into the wild where they turned feral and proliferated. As early as 1925 a Camels Destruction Act was promulgated.

BACK TO AMERICA

The south-west of the United States lacked rivers that would allow bulk transportation by steamboat, which was a major conveyor of heavy goods in the nineteenth century. Moreover, the US army found it difficult to patrol the desert lands on the country's south-western frontier. Therefore Major Henry Wayne recommended the purchase of camels to the War Department. In the Senate Jefferson Davis, the future President of the Confederate States, took up the idea as a way of strengthening the

Rival ways to cross Australia in a 1930s Trans-Australian Railway poster.

160

military presence in the region. Another senator thought that the introduction of the camel would civilize the Indians: 'For a time, to be sure, possession of this animal would perhaps only increase their powers of mischief, but it might in the long run provide the means of raising Indians to a state of semi-civilized life. Products of the camel (wool, skin and flesh) would prove of inestimable value to Indian tribes, which otherwise are fated to perish with the buffalo.' In 1855 the United States Congress allocated funds, $30,000, for the army to purchase camels in the Middle East. There were two big camel-buying expeditions. The first mission in 1856 found the Egyptians reluctant to sell, but eventually the army succeeded in acquiring 33 camels, two of which were Bactrians. The camels were disembarked in Indianola, Texas, and Arab herdsmen were imported to look after them. An initial attempt to keep the animals fenced in by an enclosure of prickly pear failed as the camels enthusiastically ate the fence. The second camel-buying expedition in the Middle East took place in 1857. But sadly, the army lacked experienced camel managers and the camels were ill-treated by their ignorant masters. Apart from using camels as pack animals, the army had hoped to use them to intimidate the Native Americans and perhaps also for a future war against the Mormons in Utah.

The American Civil War which started in 1861 put an end to the camel experiment. It had never had all that much support. People claimed that the camels panicked horses and mules, which bolted at their smell. The camels broke through or ate hedges and strayed into cultivated areas. Horse traders were hostile to the potential competition. The fact that Jefferson Davis, now the Confederate President, had originally sponsored the camel project prejudiced the Unionists against the animals. Finally, the spread of the rail network to the south-west meant that camels became irrelevant. With the ending of the experiment, camels which had escaped or been turned loose turned feral. Thereafter there were numerous strange stories about these camels. According to the story of the Red Ghost, one young army recruit was so nervous of riding a camel that they tied him to the back of a big red camel and slapped its rump. The camel galloped off and the young recruit was never seen again. Later the Red Ghost allegedly trampled a woman to death. Soon after that it was spotted with a headless rider on its back. But by the time an Arizona farmer succeeded in shooting it dead in 1893, the corpse had vanished from its back. Although Texan farmers hunted them down, feral camels were allegedly spotted as late as 1931, and even thereafter there were reported sightings of the Red Ghost and other ghost camels.[30]

WARFARE

The French conquest of Algeria began in 1830, but it was only at the opening of the twentieth century that they sought to expand their control over the southern Sahara. In 1900 the French occupied the oases of Touat. This was the key to the western Sahara and thousands of camels were used to supply the troops kept posted there. A year later the French established the Compagnies Sahariennes under the command of Captain Marie-Joseph-François-Henri Laperrine. He was famous for his austerity and endurance, and was to spend forty years in the Sahara. He was put in charge of the oases in 1901. The elite were the *méharistes*, so called because of the type of camel they rode upon. (A mahari camel is a riding camel that has been selected for speed.) The recruits were Chaambas (Berber camel nomads) with a few French officers. There were no uniforms, no stables and no barracks – just nomads in the pay of French army who moved in small groups rapidly covering large distances. Each

trooper had several camels. It was the *méharistes*, rather than the Foreign Legion, who mostly dealt with the Touareg tribesmen of the Southern Sahara and it was French policing of the desert that made the camel-borne salt caravans from the Bilma oasis south to Timbuktu possible. Since the Chaamba and the Touareg were long-standing enemies, military duty for the Chaambas was also a pleasure. It was also possible to cross the Sahara from French Algeria to French Senegal. In general, the French army was keen to preserve the traditional nomadic Touareg way of life from the encroachments of modernity. Another of the *méharistes'* jobs was to pursue Moroccans who raided for slaves. Later, in the 1960s, the *méharistes* were deployed against the FLN (the Algerian independence movement).[31]

Jean-Léon Gérôme, *Napoleon and His General Staff in Egypt*, 1867, oil on canvas.

The Gatling gun adapted to camel warfare in the 1870s in Egypt; detail from a contemporary print.

In Egypt in 1884 a camel corps, recruited from British life-guards, was founded to go the relief of General Gordon in Khartoum, but it was disbanded in the following year. Although there was no properly established camel corps thereafter until 1916, nevertheless in 1913 His Majesty's Stationery Office printed *The Camel Corps Training Provisional*. This little manual offered the following daunting morning schedule: 3 a.m. reveille, 3.30 saddle up, 4 march, 7 halt, 8 feed. There should be little halts every hour for camels to pee. Finally one should halt in daylight in order to choose good ground for barracking.

The Imperial Camel Corps was established in 1916. In the first instance its job was to protect Egypt's western frontier, but in the long term it was anticipated that camels would be needed to protect the army's right flank in its eventual advance out of Egypt into Ottoman Palestine. For its manpower it drew mostly on Australian, New Zealander and British troops who had served in Gallipoli. The Corps was based at Abbasa near Cairo. They used male camels because they were cheaper and sturdier (whereas Lawrence and his Bedouin raiders preferred the swifter and more tractable female camels). At first they were used with considerable success against the Sanussi tribesmen in the western deserts of Egypt. Whenever possible the Camel Corps in battle dismounted and fought as far as possible from their camels, as the animals presented such large and vulnerable targets. The Camel Corps fought at Gaza and Beersheba in October 1917 and then experienced heavy losses in the advance on Jerusalem.[32] According to H. M. Barker: 'Some of the Camel Corps complained that they were having to fight two wars simul-taneously, one with the Turks and the other with their camels.'[33]

After a large number of poor-quality camels had been pur-chased, Arnold Spencer Leese was put in charge of the business. He had a great deal of experience, as he had previously been

intensively researching trypanosomiasis in India. (His post-
mortem work in the corrupting heat had been regularly attended
by a circle of vultures.) In his *Treatise on the One-Humped Camel*
he stated that there were two golden rules for a camel doctor:
first, look after number one, and second, wear your oldest clothes.
He was buying camels in Somaliland and, faced with a dealer's
ring, via an interpreter, he issued the scarcely veiled threat on
behalf of the British king: 'Tell them that I am training many
young soldiers at home to accustom them to the sight of blood.'[34]
This worked. After an unsuccessful attack on Amman, the Imperial Camel Corps was slowly wound down, as the territory the
British were now fighting in was no longer suitable for camels.
The Corps was finally disbanded in 1919. The Bikaner Camel
Corps, an elite force on exceptionally large camels sent from
Rajasthan fought alongside the Imperial Camel Corps. It has

A London statue in memory of the Imperial Camel Corps soldiers killed in Egypt, Sinai and Palestine in the First World War.

been estimated that during the First World War 22,812 camels were killed on active service. A monument, sculpted by Cecil Brown and placed on London's Victoria Embankment in 1921, commemorates the human members of the Camel Corps who died in the First World War.

T. E. Lawrence praised the shock troops of the Imperial Camel Corps as a 'splendid unit of picked Yeomen and Australians', but he was critical of the way the camels were overloaded and tactically mismanaged. Though in the later stages of the war he was lent camels and men from the Camel Corps, Lawrence mostly worked with lighter and more mobile Bedouin. Lawrence set out his own view of desert warfare:

In character our operations . . . should be like naval war, their mobility, their ubiquity, their independence of bases and lack of communications . . . Camel raiding parties, self contained like ships, might cruise without danger

168

along the enemy's cultivation frontier, and tap or raid into his lines where it seemed easiest or fittest or most profitable, with always a sure retreat behind them into the desert-element which the Turks could not explore.

He added that 'the necessary speed and range at which to strike, if we were to make war in this distant fashion, would be attained through the extreme frugality of the desert men, and high efficiency when mounted on their female riding camels . . . We

T. E. Lawrence on a camel at Aqaba in Jordan, 1917, during the 'Arab Revolt'.

had found that on camels we were independent of supply for six weeks . . . Our six weeks food would give us capacity for a thousand miles out and home.'[35] The Bedouin were keen to help Lawrence in blowing up the Hejaz Railway, since, in the long run, that railway would have deprived them of the incomes they derived from demanding protection money from desert caravans and from providing camels for transport. But, on the other side of the Arabian Peninsula, Saud, who was destined to establish a kingdom over most of Arabia in the 1920s, was angry at the disruption of the hajj route and the shutdown of the camel trade from the Gulf to Syria during the War. In 1921 a conference was held in Cairo to sort out future British policy in the Middle East. An outing to the Sphinx and the Pyramids was organised for the leading delegates and they, including Winston and Clementine Churchill, Churchill's private detective, Gertrude Bell and T. E. Lawrence, posed on their camels for a photograph. Shortly thereafter, however, Churchill was thrown from his camel and injured his hand. He had to ride back with Lawrence.[36]

Winston Churchill, Gertrude Bell, T. E. Lawrence and others on camels in front of the Pyramids during the 1921 Cairo Conference, which largely created the map of the Middle East as we know it today.

A camel on Gilbert Scott's 1862–72 Albert Memorial – an emblem, presumably, of the spread of empire.

As a pendant to this account of the camel's role in First World War, Arnold Leese's subsequent career, as narrated by him in *My Irrelevant Defence, Being Meditations Inside Gaol and Out on Jewish Ritual Murder* (1938) and *Out of Step: Events in the Two Lives of an Anti-Jewish Camel Doctor* (1951), is most curious. Already during the war he had become suspicious about the prevalence of freemasonry in the barracks. But he had what he describes as his 'political wakening' in 1926. It seemed apparent to him the Jews were behind all the evil in the world. 'I think history records that England was at its best when it knew nothing of tobacco . . . and had no Jews.' He read *The Protocols of the Elders of Zion* and believed every word of that farrago. The Jew was a poison in the bloodstream of Britain, akin to trypanosomiasis. In 1928 he published his valuable treatise, *The One-Humped Camel in Health and Disease*. At some point, he also published *The Legalised Cruelty of Schechita: The Jewish Method of Cattle Slaughter* (n.d.). He also retired as a vet and became co-founder of the Imperial Fascist League. He was so far to the right that he was contemptuous of Oswald Mosley and his 'kosher fascism'. He believed that the Jack the Ripper murders and the abduction of the Lindbergh baby were cases of Jewish ritual murder. He was briefly imprisoned for libel. When the Second World War broke out, he went on the run to avoid being imprisoned under section 186, but he was overpowered and arrested in 1940. He was released in 1943 because of ill-health. However, he was re-imprisoned in 1947 for helping members of the Waffen ss to escape justice. He died in 1956. A camel parasite, *Thelagia Leesei*, was named after him. It is surely a fitting memorial.[37]

In the Second World War the camel played a diminished role. The Bikaner Camel Corps again saw service, this time in the Aden Protectorate, but since it had to have its fodder imported, the unit was expensive to maintain. Perhaps more crucially, the German army used camel trains to bring petrol to their tanks in south Russia as they pressed on in their doomed attempt to occupy the Caucasus.

7 Modernity's Camel

Around the beginning of the twenty-first century the Food and Agriculture Organization of the United Nations (FAO) estimated that there were approximately 19 million camels in the world, of which 15 million were in Africa and 4 million in Asia.[1] The African camels are concentrated in north-east Africa. There are more than five times as many dromedaries as there are Bactrians. The Bactrians, which were once spread more widely, are now chiefly found in China and Mongolia, though there are also some in Afghanistan and parts of Iran and India.

MIDDLE EAST

Although most of the dromedaries are now found in East Africa, the dromedary has become an icon of the Arab Middle East (in the same way that kangaroos are an icon of Australia or windmills of Holland). The paradox is that most modern Arabs regard the camel as a symbol of backwardness and most, though not all, Arab governments are trying to get the camel-rearing nomads to settle. The coming of the car and other economic pressures have also threatened the nomadic way of life in recent decades. Cars appeared in the Middle East after the First World War and in the early 1960s trucks started to become common in Syria and Lebanon. Syrian camel herders had already suffered in the 1930s from the Egyptian switch from buying Syrian camels for butchering to Sudanese camels, which were cheaper.

The 1958–62 drought affected the entire Middle East. Even prior to the drought, the Syrian desert had been overgrazed. Saudis reckoned that the great drought lasted from 1956 to 1972. Many of the tribesmen moved to jobs in oil or the army. The impact of drought, as well as the expansion of settled agriculture, means that herdsmen have become more dependent on permanent wells, whereas before the herds used to follow the rains. But dependence on a fixed water source tends to lead to overgrazing in the vicinity. The slow breeding rate of camels, together with a high mortality rate among calves, can swiftly bring herds to below subsistence level.

Men of the Arab Legion parade at the coronation of King Hussein of Jordan, 1953.

174

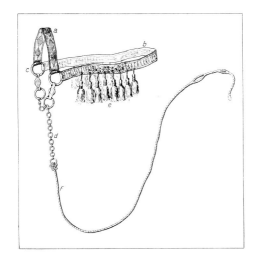

A camel harness from Alois Musil, *Manners and Customs of the Rwala Bedouin . . .* (1928).

The combination of recurrent drought and the spread of modern transport have been a disaster for the traditional way of life of the Rwala in the Syrian desert. A family used to subsist fairly comfortably on a herd of fifteen to twenty camels. The Rwala used to transport commodities across the desert, but the coming of motorized transport reduced them to a subsistence economy. During the great drought they lost about four-fifths of their herds. Not only did most camels die, but those that survived gave a reduced milk yield. Like most camel-rearing nomads, they moved around in quest of places in the desert where the rains had fallen recently. But camel herding in Syria has always been somewhat precarious, as the winters can be brutal and there is a high mortality rate among calves that are dropped then. Drought or not, sheep are safer to herd because of their faster breeding rate. Mutton fetches a better price than camel meat. Also, sheep can be shifted around in trucks. A single shepherd can mind four- to five-hundred sheep, but

Bedouin woven camel decoration, from Jordan, 1970s.

A camel and a 'hump' on a road sign in Dubai.

camel-herding is labour-intensive and a herd of forty to fifty camels will require two or maybe three men to look after them (and watering camels is particularly labour-intensive). Even so, despite the economic and social pressures, there has been some reluctance among Arab tribespeople to abandon camel-herding, because the animals are liked and they are seen as conferring political independence. Moreover, by tradition, possession of camels confers prestige.[2]

In Lebanon there were also problems with camels straying into the orchards that proliferated in the region. Camels have a marked tendency to stray while grazing and this makes them more intrusive and labour-intensive than sheep or goats. Moreover, the new paved roads provided a precarious footing for camels. The Fadl and Hassana tribes on the Syro-Lebanese border were driven out of business by the truck in the early 1960s. The herders sold off their camels and they have practically vanished from Lebanon's Bekaa valley. Many of the men went to work in the new sugar-beet factories. The laborious but sociable rituals of packing and unpacking the tent have been done away with. Women have more leisure, but they miss the solidarity of communal labour and gossip. Lack of camel hair has led to the abandonment of rug-making.[3]

Camel-raiding had until modern times been a crucial part of the way of life of the camel-rearing nomads of the Arabian Peninsula. 'Raids are our agriculture', as a saying of the Shammar tribe had it. In an important paper entitled 'Camel Raiding of the North Arabian Bedouin: A Mechanism of Ecological Adaptation', the anthropologist Louise E. Sweet argued that camel-raiding had the benign effect of distributing and redistributing camels throughout most of the Peninsula. Successful raids could compensate for loss of camels due to drought, poor pasturage or disease. Successful raiding was also a major source of prestige.[4]

Moreover, the raids were governed by conventions. Raiding was restricted to the noble tribes. Not only were women and children not to be harmed, but it was not honourable to steal all a tribe's camels. And indeed, as noted elsewhere, camel-raiding was not regarded as stealing. But Saudi Arabia banned inter-tribal raiding in 1932 and indeed the adoption of the rifle by the Bedouin had already had the effect of making such raiding more lethal and less chivalrous.

The Al Murrah, the most purely nomadic tribe in Saudi Arabia, have adapted happily to modernization. They are not a primitive people to be preserved in some kind of human zoo.

Its men find it convenient to travel long distances by truck and water can be carried on the trucks in disused oil drums. In the Arabian Peninsula sedentary semi-intensive farming of the dromedary has increased and water is often brought to the herds by water tankers. But there has also been a switch from subsistence-oriented camel herding to market-oriented sheep- and goat-herding. Improved irrigation makes non-nomadic herding easier. Easier herding means that women have more time for weaving, rather than setting up and dismantling tents.[5]

In general, attempts to promote the farming of camels for meat and milk have not been very successful. The FAO has tried to promote camel milk as source of income for Gulf states. In Saudi Arabia camel meat is the food of the urban poor and mutton the preferred meat. Most camel meat on offer is tough, because it is the old camels that get slaughtered. In 1978 the royal palace in Riyadh acquired an automatic camel milking plant. There are approximately 600,000 camels in Saudi Arabia today.

CAMEL RACING

In the Arabian Peninsula wealthy Arabs maintain prestige herds for breeding racing camels and as a gesture to the traditional way of life of the region. The subsistence herding of camels by nomads is being replaced by the rearing of camels for sport and tourism. Though camel racing is presented an Arab tradition, it is an invented tradition, for King Fahd of Saudi Arabia inaugurated such races in 1975.[6] In Saudi Arabia the King's Race is in the spring at Janadriyah racecourse near Riyadh Airport. The February meeting for camel racing at Janadariyah has expanded into a two-week cultural festival. There are approximately fifteen camel-racing tracks in the United Arab Emirates. Dubai's famous Nad al-Sheba camel racetrack was built for Sheikh Mohammed bin Rashid al-Maktoum (then Crown Prince of Dubai, now Emir of Dubai and Prime Minister of UAE). More recently, a second camel race track has opened at al-Liseli on the road to al-'Ayn. Al-'Ayn, on the border with Oman, has the largest camel market in UAE. It also has a race-track, but the camels traded at the year-long fair are bred for meat and milk. Al-'Ayn (like Bikaner in India) has been developing and marketing camel's milk ice-cream.

In 1992 it was estimated that there were approximately 14,000 racing camels and 40,000 breeding camels in the UAE. The main UAE racing season is from the end of August until mid-April, but there are races all year. Prized racing camels are fed cow's milk and honey. In the close season, their diet is alfalfa, barley and dates. Wealthy owners exercise their camels on treadmills and in swimming pools. (The treadmill allows the camel to exercise in bad weather. It allows the standardization of training protocols and allows measurement of the camel's pulse and so

A camel race in Dubai.

A 'mahari' camel treadmill made by Graber AG in Switzerland.

on. Allegedly, the camels enjoy it.) In the UAE the morning races are for princes, sheikhs and Bedouin. The afternoon races are reserved for the Bedouin only. Additionally, races are often held to celebrate marriages.

In the UAE and Saudi Arabia races can be up to ten kilometres. The stands are at the start of the racecourse, which is also its end, and most of the race is watched on closed-circuit televisions. It is difficult to marshal the camels at the starting line. The fastest can do the ten kilometres in 17.5 or 18 minutes. Usually there are about fifty camels in a race. While it is normal for camels to begin with a gallop before settling to pacing, the best camels can complete the whole race at a gallop. It is mostly females that are raced. A single rein and tapping with cane on side of the neck are used to direct the racing camel. Trainers follow the camels in trucks and communicate with the jockeys by two-way radio. Betting is banned, but there are lucrative purses. A big prize could be over a million dollars and a champion racing camel can cost over a million dollars.

In many Arab states photographing the races has been forbidden. This was due to embarrassment about the scandalous use of child jockeys. Child camel jockeys have been used in Oman, Bahrain, Kuwait, Qatar, Yemen, Sudan and Saudi Arabia. Small boys were kidnapped or purchased from parents from Pakistan, India and Bangladesh and Urdu became the language of camel racing. A Save the Children Sweden report estimated that 15,000 boy jockeys had been abducted or purchased from the Punjab alone. Boys as young as three became camel jockeys. The jockeys used to be attached to the blanket saddles by Velcro strips, but Velcro was abandoned after camel falls led to the deaths of the child jockeys, so now they have their legs hooked into ropes. But severe injuries were still common. Ideally the jockey should weigh about 25 kilograms, so the boys were starved to keep

A camel-racing robot in a Dubai paddock.

them light. Whereas the camels received high-calorie food, jockeys were given low-calorie food. Apart from racing, the boys had to groom camels and clean up the dung. Effectively this was child slavery.[7]

In 2002 Pakistan legislated against child smuggling and a few boys of Pakistani origin have been rescued and brought back to Lahore for rehabilitation.

In July 2005, under pressure from UNICEF, the UAE agreed a ban on the use of small boys. According to the decree, the jockeys must be fifteen or older and not weigh less than 45 kg. Although there are still some youthful jockeys, it is no longer forbidden to photograph camel races and in the UAE, Qatar and Kuwait human jockeys have mostly been replaced by robots. In 2005 a robot jockey, Kamel, developed in Switzerland by the robotics firm K-Team and operated by remote control, was first tried out in Qatar. The robot is a box with a little rotatable whip-arm on one side. The robot, which is about half a metre high and weighs two and a half kilos, is usually wrapped in the owner's racing colours and it is operated from a computer screen in a car which hurtles along beside the camel on the racecourse. According to Sheikh Abdullah bin Saud, the Qatari official in charge of the camel robot project 'We can't stop these races. They are part of our history and tradition, so we have tried to find an alternative.' However, it is very doubtful whether there really is much of a history of camel racing in the Gulf. It seems rather to be an invented tradition, something conjured up to link the immensely wealthy and technologically up-to-date Gulf states with a romantic but rather bogus past.[8]

The increased interest in camel-racing and its lavish patronage by princes in the Peninsula has had a knock-on effect on the sponsorship and intensification of scientific research – as can be seen in the publication of *Proceedings of the First International*

Camel Conference (1992), with papers on such matters as camel diseases, the reproductive cycle, the racing camel and the ecology of the feral camel in Australia.[9] The crown prince of Dubai, Sheikh Mohammed ibn Rashid al-Maktoum, is patron of the Camel Reproduction Centre, a scientific research establishment where the aim is to improve the breed of racing camels. Here the best racing camels are artificially inseminated, then the embryos are extracted at an early stage and planted in surrogate non-racing camels. This allows the real mother to continue racing, rather than being retired for thirteen months. Dubai has a camel hospital with hydrotherapy pools and expert vets.

TURKEY

Camel nomadism has been on the decline in Turkey throughout the twentieth century. Irfan Orga's beautiful and moving book, *The Caravan Moves On* (1958), was among other things a lament for the vanishing way of life of the mountain-dwelling Yürük:

> I must confess that the camel caravans stirred me . . . so perhaps no Turk ever loses his nomadic instincts. They spelled the last romantic survival in a commercialised world of noise and petrol-fumes. It is true that the camels were generally ugly brutes with dun-coloured hides and yellowed teeth. I think it was their unbelievable eyelashes that gave them some touching quality. They heaved and groaned, pulling back their rubbery lips in futile anger, yet when they had their proud heads erect and gazed at one mournfully, one was instantly wounded by the childish eyelashes and the look of sorrow in their eyes. Hikmet Bey said that this was moonshine. They knew neither sorrow nor joy. I clung to my belief however, and when

one of the men showed me a young camel foal and I caught its soft face in my hands, my heart was squeezed by the look of patience in its velvety eyes.[10]

Orga spent three weeks with the Yürük and was mesmerized by the tribe's total independence from and indifference to what was happening in the rest of Turkey. They were truly free.

But these days even the Yürük are switching to trucks from camels. The chief economic importance of the camel lies in the overlapping areas of camel-wrestling and tourism. Camel-wrestling is especially popular in the oxbow area where the Mediterranean and Aegean coasts meet. There is a big camel-wrestling festival at Selcuk in Turkey annually on the third

Camels fighting on a leaf from a Mughal album, 18th century.

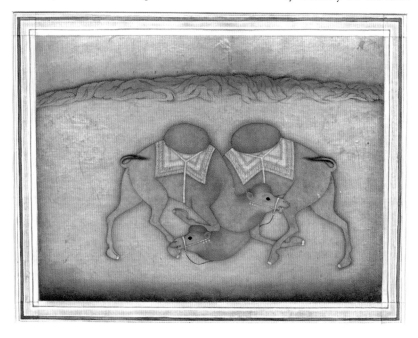

weekend in January. Elsewhere in western Turkey camel-wrestling tournaments are held on different dates, but always in the winter months, as this is the height of the rutting season. Bulls are bred and specially fed for wrestling. The wrestling camels are *tülüs* – that is, they are the result of the crossbreeding of Bactrians and dromedaries Such camels are bigger and heavier than either of their parents. At the start of a contest a female in heat is introduced to spur the two males on before being hastily led away. There are four styles, *ayak, basalti, orta* and *bas*. Fight managers like to pair camels with different fighting techniques. The camels use their necks and shove with their chests. They butt and lean against one another, in what they believe is a fight for sexual precedence. Often the fight is just a matter of half-hearted butting, but sometimes one will use his foreleg to trip his opponent. Their mouths have been tied before the match, otherwise they might bite off each other's scrota. But still lots of saliva is spewed. Bouts usually last ten to fifteen minutes and they frequently end in draws because of fears of injury to expensive fighting camels. *Urganci*, or rope workers, pull the camels apart once the bout is over. Fighting camels are elaborately caparisoned like bridegrooms and are paraded round the streets the day before the fight. There are an estimated 1,200 wrestling camels in Turkey. These days the sport is commercialized and there is a lot of betting, but apparently the sport is declining, as it is expensive to maintain a camel just for fighting.[11] Camel fights are also staged in Afghanistan.

AFRICA

Approximately 60 per cent of the world's camel population is found in East Africa. The FAO and other bodies are keen that camel rearing and herding should be encouraged in this area, as camels can be an important source of milk, much needed by the growing populations of the towns. Until the onset of repeated droughts camel herds were increasing in Somalia.

Even before the Darfur war, Sudanese herds had declined because of recurrent droughts.[12] In northern Darfur camels are herded by Arabs, but in the south cattle are preferred and settled agriculture is practised by non-Arabs. The nomadic Abbala camel-herders in northern Darfur suffered from the closing off of their traditional migratory routes by the sedentary tribes and from increased competition for sources of water. The Darfur war, which began in 2003, is in large part a war about competition for dwindling resources. Arabs from elsewhere in the Sudan have joined Arab nomads in attacking the settled African farmers and this has had covert backing from the Sudanese government. Militias on camelback fire down on village settlements. Some Abbala have joined the Janjaweed, a group whose name means 'armed and mounted outlaw'. The fighting has not all gone one way and southern villagers have often been successful in stealing camels. There used to be an interdependence between the settled and the nomadic tribes, but this has broken down and naturally the continuance of the war has made the traditional migratory routes impassable. Nomads are poorly placed, or rather not placed at all, to receive aid from aid agencies, as unlike the farmers they will not settle in displaced persons' camps.[13]

In Kenya camel pastoralism is subsistence-oriented. Here, as elsewhere in Africa, prolonged drought is killing the camels. There is a Camel Mobile Library Service for north-east Kenya's nomads.[14] Typically the camel arrives in remote regions carrying about 400 books. Pastoralists tend to drink camel's milk in a fermented state as that is the best way of preserving it. In northern Kenya, camel safaris play an important role in the tourist industry.

Mauritania is one of the countries that has made determined efforts to settle its nomads, as it regards nomadic camel herders as a threat to social stability, but there are too few opportunities for those who do settle in the rapidly growing shanty towns. Moreover, the potential resources of the desert, home to hundreds of thousands of nomads, are being lost.[15] Mauritania used to be primarily a cattle-rearing country, but drought has destroyed many herds. Although camels have also been affected by drought, their numbers have fallen less and in some parts of the country camel-rearing has replaced cattle-rearing. Even so, the camel herds are shrinking, as are the distances covered by the nomads. Traditionally the herds have moved on a north–south axis following the seasonal rains. A camels' milk dairy has been set up in the capital, Nouakchott, which pasteurizes the milk brought in by nomads. There have been experiments with 'fromage de chameau' and the FAO is assisting in this project. However, export of the cheese to Europe has proved impossible, as the European Union does not recognize the camel as a milk-producing animal.[16]

INDIA

In India the camel is widely used to pull wagons and carts, as well as to carry bales. There are between 300,000 and 450,000 camels in India. They are particularly in evidence in the desert state of Rajasthan, where there are two main types of camel – the smallish Jaisalmeri, which is primarily a riding camel, and the heavier Bikaneri, which is mostly employed in carrying heavy loads. Rajasthani camels are often decorated with a necklace known as a *gorbandh*. The great camel fair, both a serious agricultural fair and a tourist sight, is an annual event in Pushkar in November. Pushkar is a pilgrimage centre and the camel fair coincides with the festival of the Full Moon of Karvik. Pushkar is one of the world's largest livestock fairs in which thousands of camels are traded and there are other events such as a camel-loading contest to see how many riders can pile onto a loaded camel.[17] For years use of the camel for transport and farming has been in decline, but this trend seems to be about to be reversed as rising fuel prices are leading many Rajasthanis to revert to use of the camel, and the price of a camel has shot up. The proper marketing of camel's milk might also contribute to a rise in numbers.

Bikaner, in Rajasthan, furnished the famous Camel Corps that fought in the Middle East in the First World War. It still exists as a fighting and ceremonial force known as the Thirteenth Battalion of Grenadiers. Its officers enjoy camel polo. Every January there is a camel festival in Bikaner with prizes for

A camel fair in Pushkar, India.

the best types. A camel breeding farm was established there in 1960. One of its aims has been to breed camels with longer thicker eyelashes to deal with sandstorms. The Project Directorate on the Camel's research centre was established at Bikaner in 1984.[18] One finds dromedaries in Rajasthan and throughout most of India dromedaries are used. However, Bactrians are herded in the cold desert of Ladakh, so the research centre studies both. An elite herd of 270 camels has been bred. Camel ice-cream is available at its milk parlour (and is recommended).

The Raikas of Rajasthan and the Rabaris of Kutch are the traditional camel herders in India, but many are switching to sheep.[19] As in Syria, sheep herding carries less prestige, but it is more profitable. Traditional migration routes are being fenced off or reserved for military purposes. Droughts have also had an adverse effect on herds. The decline of princely India has also

Camels in the National Research Centre on the Camel, Bikaner, India.

been a factor, as the rajas used to be the patrons of prestige herds. Camels can swim and in Kutch the Rabaris swim their camels from island to island. In Uttar Pradesh thousands of camels are slaughtered every year and the meat exported to Bangladesh.

AUSTRALIA

In the 1920s, with the increased use of motorized transport in Australia, camel herds worth thousands of pounds quite suddenly became worthless and many were turned out into the wild. In 1925 the State of South Australia passed the Camel Destruction Act, which authorized the police to shoot any camels found trespassing or lacking registration disks. In 1949 the state of Western Australia (where most of the feral camels are found) passed the Vermin Act, which authorized the payment of bounty to people who killed feral camels. In recent years the state government of South Australia has culled the herds using aerial marksmen, but the culls do not seem to be having much impact. Other camels are rounded up and exported to South-East Asia, where they are slaughtered for their meat. A small amount of this meat reaches the United Kingdom, though most is sold to Saudi Arabia.

There are about a million feral camels in Australia today and their population doubles roughly every eight years. It has been estimated that Australia may one day support a population of 60 million camels. The drought in Australia in recent years has made these camels increasingly desperate and they have taken to breaking down fences, invading farms, entering houses and smashing air conditioners, pipes and lavatories in attempts to find water. Out in the desert thousands of camels are dying of thirst.[20]

There are also plenty of domesticated Bactrians around. They are kept for wool and milk. In Mongolia and China, where they are herded, the herdsman will move camp three or four times a year in search of more pasturage. Some 600,000 domesticated Bactrians are herded in the region of the Gobi Desert. But the wild Bactrians, the only wild camels in the world, are under threat of extinction. Indeed they are closer to extinction than the panda. John Hare, co-author of *The Lost Camels of Tartary*, believes that there are perhaps as few as 650 in China and 450 in Mongolia. He has set up the Wild Camel Protection Foundation in Britain.[21] The situation of the wild Bactrian is also a matter of concern to EDGE (Evolutionarily Distinct and Globally Endangered), a campaign launched by the Zoological Society of London. An international stud book for the wild Bactrian has been established.

John Hare's photograph of a wild Bactrian camel.

In China, the Lop Nor region where the wild Bactrians roamed was a nuclear test site in the years 1955–96 and in those years they were perhaps better protected from human predators. Since then illegal gold miners in the region have been slaughtering all sorts of threatened species for food. Additionally the miners leave dangerous chemicals around and sometimes they poison water sources. Some camels have been killed because they compete with farmers for pasturage. As elsewhere in the world, there has been serious drought in the area. Droughts are dangerous, not just because of less grazing becoming available, but also because there is a greater risk of camels encountering predators as the number of watering holes are reduced – in the case of the wild Bactrian camel, the danger comes from wolves. But the Bactrians in the Lop Nor region of seasonal salt-lake marshes in China are fairly safe from wolves, because only the wild Bactrians can drink the salt

A camel cartoon on an old postcard.

"Schoolboy Definitions"

The camal lives in the desart He has awlways got a hump becos he thinks about the last straw so much.

water there. The wild Bactrian calves in winter, which is, of course, exceptionally bleak in Mongolia.

The Lop Nor Wild Camel National Nature Reserve, extending over 155,000 square kilometres, has now been established. A captive breeding programme is also going on in China and Mongolia. The Wild Camel Protection Foundation (WCPF) is situated at Zakhyn Us in Mongolia in the Gobi region and is supported by the Mongolian Ministry of Nature and the Environment. The WCPF started with fifteen wild Bactrians and now has twenty-one. An area on the edge of the desert has been identified as a place in which the captive wild camels can be released into the wild. (Currently in London Zoo there are two female domesticated Bactrians, called Nadia and Noemie, and until recently they were looked after by Oliver Duprey, who had visited Zakhyn Us and joined a Mongolian herdsman's family to learn more about camel-rearing skills.) There is also a programme of radio-collaring wild Bactrians in order to track their movements. The problem though is what to do with the surplus males produced by the captive breeding programme. How can they be released into the wild?

THE FUTURE

It is difficult to make generalizations about the future of the camel. In Mongolia the wild Bactrian is threatened with extinction, but in Australia the population of feral camels continues to grow at an alarming rate. As the planet continues to heat up, camels may prosper. The use of the camel as a pack animal in commerce and warfare has obviously declined and in North and East Africa camels are mostly bred for the slaughterhouse, while in the Arabian Peninsula a new culture of camel racing has been sponsored by wealthy princes.

The famous Camel cigarette packet.

CAMEL

TURKISH & AMERICAN
BLEND
CIGARETTES

In the course of the soft-padded advance of camels across the millennia from the warm and forested plains of North America to the expanding deserts and salt-lake marshes of the modern era, though giant camels perished in prehistoric times, camels in general have grown in size. Most have formed a grudging and grumbling partnership with humans. In historic times camels have suffered a decline in prestige and dignity. In pre-Islamic Arabia they had a quasi-sacred status and were the subjects of

194

an extraordinary and extensive body of poetry. In the nineteenth century Western explorers, imperialists and soldiers such as Doughty, Palgrave, Wolseley and Kipling disparaged the animal. Citizens of Victorian Britain and Second Empire France had a cult of the horse but little time for the camel. As it declined in dignity, the camel was frequently treated unfavourably in novels. Today it is the subject of bawdy ditties and dirty jokes. Famous artists such as Gentile da Fabriano, Tiepolo and Poussin used to paint camels, but today they are more likely to be the butt of cartoonists. (Happily the camel is unaware of and indifferent to this loss of dignity.)

In the Middle Ages camels were the symbol of many things, among them humility, avarice, lust and steadfastness. Today their image is used to peddle a brand of cigarettes. They also feature as a handy shorthand icon for the Arab Middle East, though not all Arabs are happy to be represented by this image, as some see it as implying cultural backwardness. In bookshops the shelves devoted to natural history are currently crowded with books about cats, dogs, horses, tigers and whales, but not camels. Hence this book.

Timeline of the Camel

c. 40 million BC	c. 2,500,000 BC	c. 500,000 BC	c. 10,000 BC
The first camelids appear in North America	Camelids migrate into South America and reach Asia by a land bridge across the Bering Straits	The appearance of Camelus, the ancestor of the modern camel, in North America	The camel became extinct in North America

c. 550	630s onwards	656	1303–5
The *Mu'allaqa* poem of Imru' al-Qays, featuring his famous camel journey	Arab conquests in Asia and North Africa following on the defeat of Byzantine and Sassanian armies	The Battle of the Camel	Camels feature in Giotto's frescoes for the Scrovegni Chapel, Padua

1860s	1913	1916	1942
Australia begins to import camels on a large scale from the Indian subcontinent	Launch of Camel cigarettes	The British Imperial Camel Corps established in Egypt	Camels used to supply the German army's offensive in the Soviet Caucasus

c. 4000 BC	*c.* 1,000 BC	*c.* 210 BC	*c.* 150 BC
Dromedaries first domesticated in South Arabia	Bactrians first domesticated in Turkestan	Funerary ceramic camel figurines start to appear in Chinese graves	The opening up of the Silk Route

1590	*c.* 1623	1821	1850s
The camel features as the image of avarice in Spenser's *Faerie Queene*.	London's St James's Park used as a grazing ground for the royal herd of camels	Camels first seen in Japan	Camels imported by the US army on an experimental basis

1956	1958-62	2002	2005
Publication of Rose Macaulay's novel *The Towers of Trebizond* in which a camel has a leading role	Drought in the Middle East and the consequent reduction of camel herds	The first camel beauty contest held in Abu Dhabi	Robot camel jockey tried out in Qatar

References

INTRODUCTION

1 William Gifford Palgrave, *A Year's Journey through Central and Eastern Arabia (1862–63)* (London, 1883), p. 195.
2 Cf. Thomas Nagel, 'What Is It Like to Be a Bat?', *Philosophical Review*, LXXXIII (1974), pp. 435–50.
3 Ludwig Wittgenstein, *Philosophical Investigations*, trans. G.E.M. Anscombe (Oxford, 1953), p. 223.

1 PHYSIOLOGY AND PSYCHOLOGY

1 For comprehensive accounts of the anatomy and biology of the camel, see Hilde Gauthier-Pilters and Anne Innis Dagg, *The Camel: Its Evolution, Ecology, Behavior, and Relationship to Man* (Chicago, IL, 1981); R. T. Wilson, *The Camel* (London, 1984); Reuven Yagil, *The Desert Camel: Comparative Physiological Adaptation* (Basle, 1985); Andrew Higgins, *The Camel in Health and Disease* (Eastbourne, 1986).
2 William Gifford Palgrave, *A Year's Journey through Central and Eastern Arabia (1862–63)* (London, 1883), pp. 25, 26.
3 Garnet Wolseley, *The Soldier's Pocket Book*, 4th edn (London, 1882), p. 70.
4 Georg Gerster, *Sahara* (London, 1960), p. 3.
5 H. M. Barker, *Camels and the Outback* (London, 1964), p. 175.
6 Robyn Davidson, *Tracks* (London, 1980), pp. 28–9.
7 Wilfred Thesiger, *Arabian Sands* (London, 1959), p. 84.

8 A. M. Hassanein Bey, *The Lost Oases* (London, 1925), pp. 131, 134.

9 John Hare, *Shadows Across the Sahara: Travels with Camels from Lake Chad to Tripoli* (London, 2003), p. 38.

10 Richard Burton, *Personal Narrative of a Pilgrimage to Al-Madinah and Meccah*, 2 vols, 1st edn (London, 1855, Memorial edn, London 1893), vol. I, p. 234.

11 Thesiger, *Arabian Sands*, p. 83.

12 Davidson, *Tracks*, p. 196.

13 Yagil, *The Desert Camel*, pp. 45–6.

14 Gerster, *Sahara*, p. 3.

15 Tom L. Knight, *The Camel in Australia* (Melbourne, 1969), p. 109.

16 H.R.P. Dickson, *The Arab of the Desert* (London, 1949), p. 414.

17 Charles Doughty, *Arabia Deserta* [1888], 3rd edn (London, 1936), vol. II, p. 290.

18 Alexander Kinglake, *Eothen* [1844], chap. 17.

19 T. E. Lawrence, *The Seven Pillars of Wisdom: The Complete 1922 'Oxford' Text* (Fordingbridge, 2004), p. 193. (This is a fuller version of Lawrence's text than that originally made available to the general public and, in the republished form, it has an excellent index.)

20 Bertram Thomas, *Arabia Felix: Across the Empty Quarter of Arabia* (London, 1932) p. 214n.

21 F. Amin, 'The Dromedary of the Sudan', in W. Ross Cockrill, *The Camelid, An All-Purpose Animal: Proceedings of the Khartoum Workshop on Camels, December, 1971*, 2 vols (Uppsala, 1984) (twenty minute copulation), vol. I, p. 40.

22 Richard Bulliet, 'Camel', in *Encyclopaedia Iranica*, ed. Ihsan Yarshater, vol. IV (Costa Mesa, 1991), pp. 731–2.

23 Doughty, *Arabia Deserta*, vol. I, p. 369.

24 John Hare, *The Lost Camels of Tartary: A Quest into Forbidden China* (London, 1998); www.wildcamels.com/stud_book.htm.

25 Jibrail S. Jabbur, *The Bedouins and the Desert: Aspects of Nomadic Life in The East* (New York, 1996), p. 204.

26 Ibn Khaldun, *The Muqaddimah: An Introduction to History*, trans. Franz Rosenthal, 3 vols (London, 1958), vol. II, p. 348.

27 Rosita Forbes, *The Secret of the Sahara: Kufara* (London, 1921)

pp. 79–80

28 Hassanein Bey, *The Lost Oases* (London, 1925), p. 124.
29 On diseases of the camel, see Higgins, *The Camel in Health and Disease*; Peter McGrath, 'Sudan's Camel Healers', in *Dry: Life without Water*, ed. Ehsan Masood and Daniel Schaffer (Harvard, MA, 2006), pp. 16–25.
30 Lawrence, *Seven Pillars*, p. 581.
31 Hare, *Shadows*, p. 40.
32 Mark Stewart, 'Camel Evolution', *East Tennessee Creation Science Association*, III/2 (March/April 2001), p. 1.

2 ANCESTORS OF THE CAMEL

 1 On the evolution of mammals generally, see Alfred S. Romer, *The Vertebrate Story* (Chicago, IL, 1959); Sonia Cole and M. Mastland Howard, *Animal Ancestors* (London, 1964); Björn Kurtén, *The Age of Mammals* (London, 1971); Roger Osborne and Michael Benton, *The Viking Atlas of Evolution* (London, 1996); Christine M. Janis, Kathleen M. Scott and Louis L. Jacobs, eds, *Evolution of Tertiary Mammals of North America*, 2 vols (Cambridge, 1998); Donald Prothero, *After the Dinosaurs: The Age of the Mammals* (Bloomington and Indianapolis, IN, 2006).
 2 The middle Eocene is favoured by J. G. Honey, J. A. Harrison, D. R. Prothero and M. S. Stevens, 'Camelidae' in *Evolution of Tertiary Mammals of North America*, vol. I, *Terrestrial Carnivores, Ungulates, and Ungulate-Like Mammals*, ed. Janis et al., p. 439. But Prothero, *After the Dinosaurs*, pp. 190–91, has the Stenomylus camel appearing in the early Miocene.
 3 On the early camels, see Cole and Howard, *Animal Ancestors*, p. 49; Jessica A. Harrison, *Giant Camels from the Cenozoic of North America*, Smithsonian Contributions to Palaeobiology, no. 57 (Washington, DC, 1985); Honey et al., 'Camelidae'; Prothero, *After the Dinosaurs*, pp. 122, 215, 289.
 4 P. S. Martin and H. E. Wright, *Pleistocene Extinctions: The Search for a Cause*, vol. VI of *Proceedings of the VII Congress of the*

International Association for Quaternary Research (New Haven, CT, 1967): Prothero, *After the Dinosaurs*, pp. 293–5.
5 Jerome Taylor, 'Prehistoric Syrian Giant Evolved Into Modern-Day Camel', *Independent*, 9 October 2006; www.belfasttelegraph.co.uk/news/story.jsp?story=709336

3 PRACTICAL CAMEL

1 Gill Riegler, 'Buying Your First Camel', www.cameldairy.com/buying_your_first_camel.htm.
2 *Camel Corps Training Provisional 1913* (London, 1913), pp. 2–3.
3 Justin Marozzi, *South from Barbary: Along the Slave Routes of the Libyan Sahara* (London, 2001), p. 50.
4 Geoffrey Moorhouse, *The Fearful Void* (London, 1974), p. 94.
5 Sir Garnet Wolseley, *The Soldiers' Pocket Book for Field Service*, 4th edn (London, 1882), p. 70.
6 Chingiz Aitamov, *The Day Lasts More Than a Hundred Years*, trans. F. J. French (Bloomington, IN, 1983), p. 34.
7 George Gerster, *Sahara* (London, 1960), p. 5.
8 Frederick Burnaby, *A Ride to Khiva* (London, 1876), p. 213.
9 Robyn Davidson, *Tracks* (London, 1980), pp. 66–7.
10 Bimbashi McPherson, *A Life in Egypt*, ed. Barry Carman and John McPherson (London, 1983), p. 160.
11 T. E. Lawrence, *The Seven Pillars of Wisdom: The Complete 1922 'Oxford' Text* (Fordingbridge, 2004), p. 254.
12 Michael Asher, *The Last of the Bedu: In Search of the Myth* (London, 1996), p. 179.
13 André von Dumreicher, *Trackers and Smugglers in the Deserts of Egypt* (London, 1931), pp. 83–4.
14 Ibid., p. 110.
15 H.R.P. Dickson, *The Arab of the Desert: A Glimpse into Badawin Life in Kuwait and Sa'udi Arabia* (London, 1949), pp. 362–3.
16 Donald Powell Cole, *Nomads of the Nomads; the l Murrah Bedouin of the Empty Quarter* (Arlington Heights, IL, 1975), pp. 54–5.
17 Wilfred Thesiger, *Arabian Sands* (London, 1959), pp. 65–6.

18 Grenville H. Palmer, 'Sherlock Holmes in Egypt: The Methods of the Bedouin Trackers', *Strand Magazine* (August 1913), pp. 235–40. On tracking generally, see Tom Brown, *The Science and Art of Tracking* (Berkeley, CA, 1999).

19 Geert van Gelder, *Of Dishes and Discourse: Classical Arabic Literary Representations of Food* (Richmond, Surrey, 2000), p. 13.

20 Ibid., pp. 13–14.

21 *The Times*, 19 November 2007, p. 38.

22 H. M. Barker, *Camels and the Outback* (London, 1964), p. 135; Dickson, *The Arab of the Desert*, p. 416.

23 Z. Guinaudeau, *Traditional Moroccan Cooking: Recipes from Fez* (London, 1994), p. 115.

24 T. Coraghessan Boyle, *Water Music* (London, 1993), p. 54; lit.konundrum.com/features/boyletc_intv.htm.

25 Oliver Burkeman, 'Britain's Growing Taste for the Exotic', *Guardian*, 21 January 2006, p. 7.

26 *Daily Mail*, 4 February 2006, p. 7.

27 Lawrence, *Seven Pillars*, p. 299.

28 Ibn Khaldun, *Muqaddimah*, vol. I, pp. 418–19.

4 CAMELS IN THE MEDIEVAL WORLD OF ISLAM

1 On camels in the pre-modern Islamic world, see Richard W. Bulliet, *The Camel and the Wheel* (Cambridge, MA, 1975); Charles Pellat, 'Ibil', *Encyclopaedia of Islam*, 2nd edn, vol. III, p. 665; Richard Bulliet, 'Camel', in *Encylopaedia Iranica*, ed. Ehsan Yarshater (London, 1985–), vol. IV, pp. 730–33.

2 Al-Damiri, *Hayat al-Hayawan (a Zoological Lexicon)*, trans. A.S.G. Jaykar (London, 1906), p. 32.

3 Jahiz, *Le Cadi et la Mouche: anthologie du livre des animaux*, ed. and trans. Lakhdar Souami (Paris, 1988), p. 356; cf. Annemarie Schimmel, *The Triumphal Sun: A Study of the Works of Jalaloddin Rumi* (London and the Hague, 1978), pp. 122–3.

4 William Robertson Smith, *Lectures on the Religion of the Semites*, 2nd edn (London, 1894), p. 338.

5 Mircea Eliade, *Occultism, Witchcraft and Cultural Fashions: Essays in Comparative Religions* (Chicago, IL, 1976), pp. 6–8, cf. Joseph Henninger, 'Ist der sogennante Nilus-Bericht eine brauchbare religionsgeschichtliche Quelle?', *Anthropos*, L (1955), pp. 81–148.

6 Edward William Lane, *An Arabic-English Lexicon* (London, 1863–93, repr. in 2 vols, Cambridge, 1984), vol. II, pp. 2107–8.

7 R. A. Nicholson, *A Literary History of the Arabs* (London, 1907), pp. 85–6; *Encyclopaedia of Islam*, 2nd edn, ed. H.A.R. Gibb et al., 13 vols (Leiden, 1960–2009), vol. III, pp. 274–5, s.v.; Hatim al-Ta'i (C. van Arendonk); Ulrich Marzolph and Richard van Leeuwen, *The Arabian Nights Encyclopedia*, vol. I, pp. 216–17.

8 Charles Lyall, *Mufadaliyyat: An Anthology of Ancient Arab Odes*, 2 vols (Oxford, 1918), p. 28n; T. Fahd, '*La Divination Arabe: études religeuses, sociologiques et folkloriques sur le milieu natif de l'Islam*, (Leiden, 1966), pp. 204–13; T. Fahd, 'Maysir', in *Encyclopaedia of Islam*, vol. VI, pp. 923–4.

9 Charles James Lyall, *Translations of Ancient Arabic Poetry, Chiefly Pre-Islamic* (London, 1930), p. xxv.

10 On the pre-Islamic *qasida*, see Nicholson, *A Literary History*, pp. 77–9; Renate Jacobi, 'The Camel Section of the Panegyrical Ode' *Journal of Arabic Literature*, XIII (1982), pp. 1–22; Suzanne Pinckney Stetkevych, *The Mute Immortals Speak: Pre-Islamic Poetry and the Poetics of Ritual* (Ithaca and London, 1983), pp. 3–54; Stefan Sperl and Christopher Shackle, eds, *Qasida Poetry in Islamic Asia and Africa*, 2 vols, (Leiden, 1996); Robert Irwin, *Night and Horses and the Desert: An Anthology of Classical Arabic Literature* (London, 1999), pp. 3–7.

11 A. J. Arberry, *The Seven Odes* (London, 1957), p. 61.

12 On Tarafa and his ode, see Michael Sells, ' 'The Mu'allaqa of Tarafa', *Journal of Arabic Literature*, XVII (1986), pp. 21–33; Thomas Bauer, 'Tarafa ibn al-'Abd', in Julie Scott Meisami and Paul Starkey, *Encyclopedia of Arabic Literature* (London and New York, 1998), vol. II, p. 759.

13 Arberry, *Seven Odes*, p. 83.

14 Sir William Jones, cited in ibid., p. 78.

15 W. Heinrichs, 'Radjaz', *Encyclopedia of Islam* (2-vol edn), vol. VIII, pp. 375–9.

16 Richard Burton, *Personal Narrative of a Pilgrimage to Al-Madinah and Meccah*, 2 vols, 1st edn (London, 1855, Memorial edn, London 1893), vol. II, p. 133n.

17 Joseph von Hammer-Purgstall, 'Das Kamel', *Kaiserliche Akademie der Wissenschaften zu Wien, Philosophisch-historische Classe. Denkschriften*, vol. VI (1855), pp. 1–84, vol. VII (1856), pp. 1–104; cf. Georg Krotkoff, 'Hammer-Purgstall's Schrift "Das Kamel"', *Wiener Zeitschrift fur die Kunde des Morgenlands*, LXXXII (1992), pp. 261–8.

18 Wilfred Thesiger, *Arabian Sands* (London, 1959), p.84.

19 Lane, *An Arabic-English Lexicon*.

20 Jorge Luis Borges, 'The Argentine Writer and Tradition', in Borges, *Selected Non-Fictions*, ed. Eliot Weinberger (New York, 1999), pp. 423–4.

21 Barnaby Rogerson, *The Heirs of the Prophet Muhammad and the Roots of the Sunni-Shia Schism* (London, 2006) p. 296; cf. on the Battle of the Camel, L. Veccia Vaglieri, 'al-Djamal', in *The Encyclopaedia of Islam* (2-vol edn), vol. I, p. 414; G.R. Hawting, *The First Dynasty of Islam*, (Beckenham, Kent, 1986), p. 27.

22 Al-Asma'i, *Kitab al-Ibil*, ed. A. Haffner (Baghdad, 1905); Herbert Eisenstein, *Einführung in Arabische Zoographie* (Berlin, 1991), p. 55; M. G. Carter, 'Al-Asma'i', in *Routledge Encyclopedia*, vol. I, p. 110; B. Lewin, 'Al-Asma'i', *Encyclopedia of Islam* (2-vol edn), vol. I, p. 717.

23 Al-Jahiz's *Book of Beasts* exists in an Arabic edition, *Kitab al-Hayawan*, ed. 'Abd al-Salam M. Harun, 8 vols, 2nd edn (Cairo, 1965–7). Selected extracts have been published in French as Jahiz, *Le Cadi et la Mouche: anthologie du Livre des Animaux*, ed. and trans. Lakhdar Souami (Paris, 1988) and Al-Jahiz, *Le Livre des animaux: de l'étonnante sagesse divine dans sa création et autres anecdotes*, trans. Mohamed Mestiri (Paris, 2003). On the *Book of Beasts*, see N. Bel-Haj Mahmoud, *La Psychologie des Animaux chez les Arabes, notamment à travers le Kitab al-Hayawan de Dhahiz* (Paris, 1977). On the fascinating figure of al-Jahiz more generally, see Charles Pellat, *Le Milieu Basrien et la Formation de Jahiz* (Paris,

1953); Charles Pellat, ed., *The Life and Works of Jahiz*, trans. D. M. Hawke (Berkeley and Los Angeles, 1969); Donald Richards, 'Al-Jahiz' in Meisami and Starkey, *Encyclopedia*, vol. I, pp. 408–9.

24 Al-Jahiz, *Le Cadi et la Mouche*, p. 276.

25 Joel. L. Kraemer, *Humanism in the Renaissance of Islam: The Cultural Revival during the Buyid Age* (Leiden, 1993), p. 262.

26 Len E. Goodman, *The Case of the Animals Versus Man Before the King of the Jinn: A Tenth-century Ecological Fable of the Pure Brethren of Basra* (Boston, MA, 1978), pp. 60–61.

27 Van Gelder, *Of Dishes and Discourse*, pp. 12–13.

28 R. A. Nicholson, *Translations of Eastern Poetry and Prose* (Cambridge, 1922), p. 104.

29 Al-Damiri, *Hayat al-Hayawan (A Zoological Lexicon)*, trans. A.S.G. Jaykar, 2 vols (London, 1906), vol. I, pp. 26–34, 434–47.

30 Burton, *The Arabian Nights*, vol. VI, p. 92n.

31 Al-Damiri, *Hayat al-Hayawan*, vol. I, p. 447.

32 Richard Bulliet, *The Camel and the Wheel* (Cambridge, MA, 1975), p. 220.

33 Adel T. Adamova and Michael J. Rogers, 'The Iconography of a Camel Fight', *Muqarnas*, XXI, pp. 1–14.

5 THE BEAUTY OF THE BEAST: LITERATURE AND ART

1 Timon Screech, *Sex and the Floating World: Erotic Images in Japan* (London, 1999), pp. 159–60.

2 Tetsuo Nishio, 'The *Arabian Nights* and Orientalism from a Japanese Perspective', in Yuriko Yamanaka, *The Arabian Nights and Orientalism: Perspectives from East and West* (London, 2006), p. 157.

3 Elfriede R. Knauer, *The Camel's Load in Life and Death: Iconography and Ideology in Chinese Pottery* (Zurich, 1998).

4 D.D.R. Owen, trans., *The Romance of Reynard the Fox* (Oxford, 1994), pp. 94–5.

5 John Ruskin, *The Bible of Amiens* (London, 1885), in *The Works of John Ruskin*, ed. E. T. Cook and Alexander Wedderburn, 39 vols (London and New York, 1903–12), vol. XXXIII, p. 153.

6 Richard Ettinghausen, Oleg Grabar, Marilyn Jenkins-Madina, *Islamic Art and Architecture 650–1250* (New Haven and London, 2001), p. 299; Oleg Grabar, 'The Experience of Islamic Art', in *The Experience of Islamic art on the Margins of Islam*, ed. Irene A. Bierman (Reading, 2005), pp. 30–47.

7 Beryl Rowland, *Animals with Human Faces: A Guide to Animal Symbolism* (London, 1974), pp. 48–50.

8 Geoffrey Chaucer, 'The Pardoner's Tale', in *The Canterbury Tales*, in *The Riverside Chaucer*, ed. Larry D. Benson, 3rd edn (Oxford, 1988); Rowland, *Animals with Human Faces*, p. 49.

9 Edmund Spenser, *The Faerie Queen*, I, iv, 27.

10 Michael Levey, *Giambattista Tiepolo: His Life and Art* (New Haven, CT, and London, 1986), p. 198; Svetlana Alpers and Michael Baxandall, *Tiepolo and the Pictorial Intelligence* (New Haven, CT, and London), p. 162.

11 Thomas L. Glen, 'A Note on Rebecca and Eliezer at the Well of 1648', *Art Bulletin*, LVII (June 1955), pp. 221–4; Moshe Barasch, *Theories of Art*, vol. I, *From Plato to Winkelmann* (London, 2000), pp. 366–7; Paul Smith and Carolyn Wilde, *A Companion to Art Theory* (Oxford, 2002), p. 96; Christopher G. Hughes, 'Embarras and Discovenance in Poussin's *Rebecca and Eliezer at the Well*', *Art History*, XXIV (2003), pp. 493–519.

12 Alastair Hamilton, *Arab and Islamic Culture in the Heritage Library of Qatar Foundation: The European Legacy* (n.p., 2006), p. 61; Sarah Wimbush, 'Elijah Walton (1832–1880), Landscape Painter' in *Oxford Dictionary of National Biography*, ed. H.G.C. Matthew and Brian Harrison, 60 vols (Oxford, 2004), vol. 57, pp. 198–9.

13 Kristian Davies, *The Orientalists: Western Artists in Arabia, Persia and India* (New York, 2005) p. 56.

14 A monograph on John Frederick Lewis by Emily Weeks is forthcoming.

15 John Ruskin, 'Academy Notes 1856', in *The Works*, vol. XIV, p. 74.

16 Ruskin, 'Academy Notes 1859', in *The Works*, vol. XIV, p. 219.

17 Robert Burton, *The Anatomy of Melancholy* [1651], ed. Holbrook Jackson (London, 1932), book 3, section 3, p. 262.

18 Sir Thomas Browne, *Religio Medici*, in *The Works of Sir Thomas Browne*, ed. Charles Sayle, 3 vols (Edinburgh, 1912), vol. I, p. 24.

19 Gustave Flaubert, *Madame Bovary*, trans. Alan Russell (Harmondsworth, 1950), p. 242.

20 Francis Steegmuller, trans. and ed., *Flaubert in Egypt* (London, 1972), p. 43; cf. Geoffrey Wall, *Flaubert: A Life* (London, 2001), p. 177

21 Julian Barnes, *Flaubert's Parrot* (London, 1984), pp. 54–5.

22 Rose Macaulay, *The Towers of Trebizond* (London, 1956), p. 7.

23 John F. Keane, *Six Months in the Hejaz* (London, 1887), pp. 362–3.

24 Robyn Davidson, *Tracks* (London, 1980), p. 65.

25 At en.wikipedia.org/wiki/Mankurt.

26 Chingiz Aitamov, *The Day Lasts Longer Than a Hundred Years*, (London, 1983); cf. John Clute and Peter Nicholls, eds., *The Encyclopedia of Science Fiction* (London, 1993), pp. 9–10.

27 Aitamov, *The Day*, p. 41.

28 Ibid., pp. 173–4.

29 Ibid., p. 280.

30 Rudyard Kipling, *Rudyard Kipling's Verse: Definitive Edition* (London, 1943), pp. 408–9.

31 Ibid., p. 247.

32 Rudyard Kipling, *Just So Stories* (London, 1902), pp. 15–27.

33 Charles Doughty, *Arabia Deserta* [1888], 3rd edn (London, 1936), vol. I, p. 262.

34 Ibid., vol. I, p. 472.

35 T. E. Lawrence, *The Seven Pillars of Wisdom: The Complete 1922 'Oxford' Text* (Fordingbridge, 2004), p. 187.

36 Ibid., p. 300.

37 Sam Cottingham Rolls, *Steel Chariots in the Desert* (London, 1937).

38 Howard Kent, *Single Bed for Three: A Lawrence of Arabia Notebook* (London, 1963), p. 139. On the filming of Lawrence more generally see Adrian Turner, *The Making of David Lean's Lawrence of Arabia* (Limpsfield, Surrey, 1994); Steven C. Caton, *Lawrence of Arabia: A Film's Anthropology* (Berkeley, CA, 1999); Kevin Jackson, *Lawrence of Arabia* (London, 2007).

39 Kent, *Single Bed for Three*, p. 103.

1 William Foxwell Albright, *The Archaeology of Palestine* (Harmondsworth, 1949), p. 106.

2 R. T. Wilson, *The Camel* (London, 1984), pp. 5–7.

3 The fullest discussion of the chronology of domestication is to be found in Richard W. Bulliet, *The Camel and the Wheel* (Cambridge, MA, 1975). See also reviews of this book by R. B. Sergeant in *Bulletin of the School of Oriental and African Studies*, XXXIX (1976), pp. 662–4 and by Peter von Sivers in *Journal of the American Oriental Society*, IX (1978), pp. 163–4.

4 On the domestication of the Bactrian and the silk route, see Jean Paul Roux, 'Le Chameau en Asie Centrale: son nom – son elevage – sa place dans la mythologie', *Central Asiatic Journal*, V (1959), pp. 35–76; Bulliet, *The Camel and the Wheel*, pp. 148–56, 164, 170; Frances Wood, *The Silk Road: Two Thousand Years in the Heart of Asia* (London, 2002); E. E. Kuzmina, *The Prehistory of the Silk Road* (London, 2008).

5 Bulliet, *The Camel and the Wheel*, pp. 77–86.

6 Herodotus, *The Histories*, trans. Aubrey de Sélincourt, (Harmondsworth, 1954), p. 218.

7 Bulliet, *The Camel and the Wheel*, p. 107.

8 Juliet Clutton-Brock, *A Natural History of Domesticated Mammals*, 2nd edn (Cambridge, 1999), p. 159.

9 E. W. Bovill, *The Golden Trade of the Moors*, 2nd edn (Oxford, 1968); Bulliet, *The Camel and the Wheel*, pp. 111–40; Bulliet, 'Botr et Beramés: hypothèses sur l'histoire des Berbères, *Annales: economies sociétés, civilisations*, XXXVI (1981), pp. 104–16.

10 Bulliet, *The Camel and the Wheel*, pp. 92–3, 100–2, 108–9; Christian Augé and Jean-Marie Dentzer, *Petra, The Rose-Red City* (London, 2000), p. 31. 'The Camel, Ancient Ship of the Desert and the Nabataeans', nabataea.net/camel.htmml.

11 Bulliet, *The Camel and the Wheel*, p. 91.

12 Ibid., pp. 19–21, 101–4, 106–9.

13 Thesiger, *Arabian Sands* (London, 1959), pp. 57–8.

14 Bulliet, *The Camel and the Wheel*, especially pp. 216–36.

15 Crone 'Quraysh and the Roman Army: Making Sense of the Meccan Leather Trade', *Bulletin of the School of Oriental and African Studies*, LXX (2007), pp. 63–88.

16 D. R. Hill, 'The Role of the Camel and the Horse in the Early Arab Conquests', in *War, Technology and Society in the Middle East*, ed. V. J. Parry and M. E. Yapp (London, 1975), pp. 32–43.

17 Adam J. Silverstein, *Postal Systems in the Pre-Modern Islamic World*, (Cambridge, 2007), pp. 112–13.

18 David Ayalon, 'The System of Payment in Mamluk Military Society', *Journal of Economic and Social History of the Orient*, I (1958), pp. 270–71.

19 Ahmad Darrag, *L'Égypte sous le Règne de Barsbay* (Damascus, 1961), p. 90.

20 'The Travels of Bertandon de la Brocquière' in Thomas Wright, *Early Travels in Palestine* (London, 1848), pp. 300–1.

21 Simone Sigoli, *Viaggio al-Monte Sinai* (Milan, 1841), p. 80.

22 Felix Fabri, *Le Voyage en Egypte de Felix Fabri, 1483*, ed. and trans. Jacques Masson, 3 vols (Paris, 1975), vol. I, pp. 48–56.

23 Arnold von Harff, *The Pilgrimage of Arnold Von Harff Knight*, ed. and trans. Malcolm Letts (London, 1946), pp. 134–6.

24 Bulliet, *The Camel and the Wheel*, p. 241.

25 Ibid., pp. 240–41.

26 Samuel C. Chew, *The Crescent and the Rose: Islam and England during the Renaissance* (New York, 1937), p. 468.

27 Geoffrey Parker, *Philip II* (Chicago, 1995), pp. 40–41.

28 Bulliet, *The Camel and the Wheel*, pp. 241–2.

29 H. M. Barker, *Camels and the Outback* (London, 1964); see also Tom L. Knight, *The Camel in Australia* (Melbourne, 1969); Kathy Marks, 'Australia Remembers the "Cameleers"', *Independent*, 14 December 2007, pp. 44–5.

30 *Report of the Secretary of State of War Communicating in Compliance with a resolution of the senate of February 2 1857. Information Respecting the Purchase of Camels for the purposes of military Transportation* (Washington, 1857); Odie B. Faulk, *The US Camel*

Corps: An Army Experiment (New York, 1976). Joe Zentner, 'The Great Southwestern Desert Camel Experiment', www.desertusa.com/mag05/sep/camel.html.

31 On the French in North Africa, E.W. Bovill, *The Golden Trade of the Moors*, 2nd edn (Oxford, 1968), p. 250; Hilde Gautier-Pilters and Anne Innis Dagg, *The Camel: Its Evolution, Ecology and Relationship to Man* (Chicago, 1981), pp. 156–61.

32 On the British Camel Corps, see *Camel Corps Training Provisional*, (London, 1913); G. Keays, 'A Note on the History of the Camel Corps', *Sudan Notes and Records*, xxii (1939), pp. 135–7; C. S. Jarvis, *Arab Command: The Biography of Lieutenant-Colonel F. W. Peake Pasha;* Geoffrey Inchbald, *Camels and Others* (London, 1968) and *Imperial Camel Corps* (London, 1970); Neville Rowen Forth, *A Fighting Colonel of Camel Corps: The Life and Experience of Lt. Col. N. B. de Lancey Forth DSO (and bar) MG, 1879–1933 of the Manchester Regiment and Egyptian Army* (Bramton, Devon, 1991). See also Juliet Gardiner, *The Animals' War: Animals in Wartime from the First World War to the Present Day* (London, 2006), pp. 55–67.

33 Barker, *Camels and the Outback*, p. 170.

34 Arnold Leese, *Out of Step: Events in the Two Lives of an Anti-Jewish Camel-doctor* (Guildford, 1951).

35 Wilson, *The Camel*, pp. 425–6; T. E. Lawrence, *Seven Pillars of Wisdom: The Complete 1922 'Oxford' Text* (Fordingbridge, Hampshire, 2004), pp. 368–9.

36 Jeremy Wilson, *T. E. Lawrence: Lawrence of Arabia* (London, 1988), p. 139.

37 Arnold Leese, *The One-Humped Camel in Health and in Disease*, (Stamford, CA, 1927); 'Arnold Spencer Leese', at http://en.wikipedia.org/wiki/Arnold_Spencer_Leese.

7 MODERNITY'S CAMEL

1 Z. Farah and A. Fischer, *Milk and Meat from the Camel: Handbook on Products and Processing* (Zurich, 2004), p. 15; cf. R. T. Wilson, *The Camel* (London, 1984), pp. 18–19.

2 William Lancaster, *The Rwala Bedouin Today* (Cambridge, 1981).

3 Dawn Chatty, 'The Pastoral Family and the Truck', in Philip Carl Salzman, *When Nomads Settle: Processes of Sedentarization as Adaptation and Response* (New York, 1980), pp. 80–94.

4 Louise E. Sweet, 'Camel Raiding of the North Arabian Bedouin: A Mechanism of Ecological Adaptation', *American Anthropology*, LXVII (1965), pp. 1132–50.

5 Donald Powell Cole, *Nomads of the Nomads: The al Murrah Bedouin of the Empty Quarter* (Arilington Heights, IL, 1975).

6 On camel racing in the Arabian Peninsula, see D. H. Snow, 'An Introduction to the Racing Camel', in W. J. Allen et al., *Proceedings of the First International Camel Conference* (Newmarket, Suffolk, 1992), pp. 215–17; R. J. Rose, D. L. Evans, P. K. Knight, P. Henckel, D. Cluer and B. Saltin, 'Muscle Fibre Types: Fibre Recruitment and Oxygen Uptake During Exercise in the Racing Camel', in ibid., pp. 219–22; W. R. Cook, 'Some Observations on Respiration in the Racing Camel', in ibid., pp. 235–42; J. Kohnke and D. Cluer, 'Practical Feeding and Nutrition of Racing Camels: A Preliminary Survey' in ibid., pp. 247–50.

7 *Time Out Dubai, Abu Dhabi and the UAE* (London, 2007), p. 218; *USA Today*, 20 April 2005, p. 10A; en.wikipedia.org/wiki/Robot_jockey; 'Photo in the News: Robot Jockeys Race in Qatar', *National Geographic News*, 9 December 2008, news.nationalgeographic.com/news/2005/07/0715_robot_jockey.html.

8 Sulayman Khalaf, 'Poetics and Politics of Newly Invented Traditions in the Gulf: Camel Racing in the United Arab Emirates', *Ethnology*, XXXIX (2000), pp. 243–61.

9 Allen et al., *Proceedings of the First International Camel Conference*.

10 Irfan Orga, *The Caravan Moves On* (London, 1958), pp. 58–9.

11 'Camel Wrestling', en.wikipedia.org/wiki/Camel_wrestling; 'Camel wrestling', www.allaboutturkey.com/camel.htm.

12 Michael Asher, *A Desert Dies* (Harmondsworth, 1986).

13 'A Leading Player in the Darfur Drama: The Hapless Camel, *NY Times News Service, Fagoo, Sudan*, 6 December 2005, p.6. On the fighting in Darfur more generally, see M. W. Daly, *Darfur's*

Sorrow: A History of Destruction and Genocide (New York, 2007); Richard Rottenburg, *Nomadic-Sedentary Relations and Failing State Institutions in Darfur and Kordofan (Sudan)* (Halle, 2008).

14 *The Observer*, 4 December 2005, p.13. There is a novel on the subject: Masha Hamilton, *The Camel Bookmobile* (London, 2007). Book donations can be sent to Garissa Provincial Library, For Camel Library, Provincial Librarian, Rashid M. Farah, P.O. Box 245–70100, Garissa, Kenya.

15 Hilde Gautier-Pilters and Anne Innis Dagg, *The Camel: Its Evolution, Ecology and Relationship to Man* (Chicago, 1981), pp. 143–5 and 162.

16 'Why Camelbert Cheese May Give Europe the Hump', *Independent*, 27 April 2007, p. 33; 'Mauritania Camel Dairy Pioneer Tells of Changes', *The Times*, 21 April 2006, p. 17.

17 Abigail Hole, Martin Robinson, Sarina Singh, *Rajasthan, Delhi and Agra: Lonely Planet* (London, 2005), p. 212; *The Times*, 16 November 2005, p. 37.

18 'National Research Center on the Camel', at www.icar.org.in/nrccm/index.php. The website of the Bikaner-based *Journal of Camel Practice and Research* is www.camelsandcamelids.com.

19 Robyn Davidson, *Desert Places* (London, 1996); 'Rajasthan chief minister commits to camels', www.lpps.org/.

20 Bernard Langan, 'Camels go on Rampage After they get Hump Over Drought', *The Times*, 9 December 2006, p. 54; James Woodford, 'Welcome to Camel Country: Future Population 60 Million', www.smh.com.au/articles/2002/12/20/1040174390961.html. The website of *Australian Camel News, Racing and Information* is at www.austcamel.au/informn.htm

21 John Hare, *The Lost Camels of Tartary: A Quest into Forbidden China* (London, 1998). The Wild Camel Protection Foundation is on the Web at www.wildcamels.com. See also Matthew Parris, 'Desert Downturn', *The Times*, 23 October 2008, p. 28.

Select Bibliography

Adamova, Adel, and M. J. Rogers, 'The Iconography of a Camel
    Fight', *Muqarnas*, XXI (2004), pp. 1–14
Aitmatov, Chingiz, *The Day Lasts More Than a Hundred Years*, trans.
    John French [1980] (London, 1983)
Allen, W. J., et al, *Proceedings of the First International Camel
    Conference* (Newmarket, Suffolk, 1992)
Arberry, A. J., *The Seven Odes* (London, 1957)
Asher, Michael, *A Desert Dies* (Harmondsworth, 1986)
—, *The Last of the Bedu: In Search of the Myth* (London, 1996)
Barker, H. M., *Camels and the Outback* (London, 1964)
Bel-Haj Mahmoud, N., *La psychologie des des animaux chez les Arabes*
    (n.p., 1977)
Bulliet, Richard W., *The Camel and the Wheel* (Cambridge, MA, 1975)
—, *Hunters, Herders and Hamburgers: The Past and Future of Human-
    animal Relationships* (New York, 2005)
—, 'Botr et Beranés: hypotheses sur l'histoire des Berbères', *Annales,
    economies, sociétés, civilisations*, XXXVI (1981), pp. 104–16
—, *Encyclopedia Iranica*, s.v. 'Camel'
Burton, Sir Richard Francis, *First Footsteps in Africa; Or, an
    Exploration of Harar* [1856] (London, 1966)
Chenety, David R., 'The Meaning of the Cloud in Hamlet',
    Shakespeare Quarterly, X (1959), pp. 446–7
Chew, Samuel, C., *The Crescent and the Rose: Islam and England During
    the Renaissance* (New York, 1937)
Cloudsley-Thompson, John, *Deserts and Grasslands: The World's Open*

Spaces (London, 1975)

Cockrill, W. S., ed., *The Camelid: An All-Purpose Animal*, 2 vols (Uppsala, 1984)

Cole, Donald Powell, *Nomads of the Nomads: The ᴦ Murrah Bedouin of the Empty Quarter*, (Arlington Heights, IL, 1975)

Cole, Sonia and Maitland Howard, *Animal Ancestors* (London, 1964)

Crone, Patricia, *The Meccan Trade and the Rise of Islam* (Oxford, 1987)

—, 'Quraysh and the Roman Army: Making Sense of the Meccan Leather Trade', *Bulletin of the School of Oriental and African Studies*, LXX (2007), pp. 63–88

Al-Damiri, *Hayat al-Hayawan (a Zoological Lexicon)*, trans. A.S.G. Jaykar, 2 vols (London, 1906)

Davidson, Robyn, *Desert Places* (London, 1996)

—, *Tracks* (London, 1980)

Davies Kristian, *The Orientalists: Western Artists in Arabia Persia and India* (New York, 2005)

Davies, Reginald, *The Camel's Back: Service in the Rural Sudan* (London, 1957)

Dickson, H.R.P., *The Arab of the Desert* (London, 1949)

Donlon, John G., 'Camel Wrestling', *Bizarre*, XL (n.d.), pp. 87–9

Doughty, Charles, *Travels in Arabia Deserta*, 2 vols. (Cambridge, 1888)

Dumreicher, André von, *Trackers and Smugglers in the Deserts of Egypt* (London, 1931)

Eisenstein, *Einfuhrung in Arabische Zoographie: das tierkundliche Wissenschaft in der arabisch-islamisch Literatur* (Berlin, 1991)

Eliade, Mircea, *Occultism, Witchcraft and Cultural Fashions: Essays in Comparative Religions* (Chicago, IL, 1976)

Fahd T., s.v. 'al-Maysir', in *Encyclopaedia of Islam* (2nd edn)

Farah, Z. and A. Fischer, eds, *Milk and Meat from the Camel: Handbook on Products and Processing* (Zurich, 2004)

Faulk, B. Odie, *The US Camel Corps: An Army Experiment* (New York, 1976)

Fergus Fleming, *The Sword and the Cross* (London, 2003)

Foltz, Richard, C., *Animals in Islamic Tradition and Muslim Cultures* (Oxford, 2006)

Forbes, Rosita, *The Secret of the Sahara: Kufara* (London, 1921)

Forth, Nevill de Rouen, *A Fighting Colonel of the Camel Corps. The Life and Experiences of Lt. Col. N.B. DSSO (& Bar), MG 1879–1933 of the Manchester Regiment and Egyptian Army* (Braunton, Devon, 1991)

Freud, Sigmund, *Totem and Taboo*, trans. James Strachey (London, 1950)

Gardiner, Juliet, *The Animal's War: Animals in Wartime from the First World War to the Present Day* (London, 2006)

Gautier-Pilters, Hilde, and Anne Innis Dagg, *The Camel: Its Evolution, Ecology, Behaviour and Relationship to Man* (Chicago, 1981)

Haines, Tim and Paul Chambers, *The Complete Guide to Prehistoric Life* (London, 2005)

Hare, John, *The Lost Camels of Tartary: A Quest into Forbidden China* (London, 1998)

—, *Lost Cities in the Heart of Asia* (forthcoming, 2009)

—, *Shadows Across the Sahara: Travels with Camels from Lake Chad to Tripoli*, (London, 2003)

Harrison, Jessica, A., *Giant Camels from the Cenozoic of North America*, (Smithsonian Contributions to Paleobiology, no. 57) (Washington, 1985)

Higgins, Andrew, ed., *The Camel in Health and Disease* (Eastbourne, 1986).

Hill, D. R., 'The Role of the Camel and the Horse in the Early Arab Conquests', in *War, Technology and Society in the Middle East*, ed. V. J. Parry and M. E. Yapp (London, 1975), pp. 32–43

Inchbald, Geoffrey, *Camels and Others* (London, 1968)

—, *Imperial Camel Corps* (London, 1970)

Jackson, Clarence, J.-L., (a.k.a. Bulliet), *Kicked to Death by a Camel* (New York, 1973)

Ingham, Bruce, 'Camel Terminology Among the Al Murrah Bedouins', *Zeitschrift für arabische Linguistik*, XXII (1990), pp. 67–78

Jabbur, Jibrail S., *The Bedouins and the Desert: Aspects of Nomadic Life in the Desert*, trans. Lawrence Conrad (New York, 1996)

Jacobi, Renata, 'The Camel Section of the Panegyrical Ode', *Journal of Arabic Literature*, XIII (1982), pp. 1–22

Jarvis, C. S., *Arab Command: The Biography of Lieutenant-Colonel F. W. Peake Pasha* (London, 1942).

—, *Yesterday and Today in Sinai (Edinburgh, 1931)*

Jones, Alan, *Early Arabic Poetry Volume One: Marathi and Sul'uk Poems* (Reading, 1992)

—, *Early Arabic Poetry Volume Two: Select Odes* (Reading, 1996)

Kaye, Alan S., 'Semantic Transparency and Number Marking in Arabic and Other Languages' *Journal of Semitic Studies*, L (2005), pp. 155–96.

Keane, John F., *Six Months in the Hejaz: An Account of the Mohammedan Pilgrimage to Mecca and Medinah: Accomplished by an Englishman Professing Mohammedanism* (London, 1887)

Kent, Howard, *Single Bed For Three: A 'Lawrence of Arabia' Notebook*, (London, 1963)

Khalaf, Sulayman, 'Poetics and Politics of Newly Invented Traditions in the Gulf; Camel Racing in the United Arab Emirates', *Ethnology*, XXXIX (2000), pp. 243–61

Kingdon, Jonathan, *Arabian Mammals: A Natural History* (London, 1991)

Kinglake, Alexander, *Eothen* (London, 1844)

Kipling, John Lockwood, *Beast and Man in India* (London, 1891)

Knauer, Elfriede Regina, *The Camel's Load in Life and Death* (Zurich, 1998)

Knight, Tom L., *The Camel in Australia* (Melbourne, 1969)

Konigsberger, E. L., *Journey by First Class Camel* (London, 1983)

Kurtén, Björn, *The Age of Mammals* (London, 1971)

Kuzmina, E. E. *The Prehistory of the Silk Road* (Philadelphia, PA, 2008)

Lancaster, William, *The Rwala Bedouin Today* (Cambridge, 1981)

Lane, Edward William, *An Arabic-English Lexicon* [1863–93], 2 vols (Cambridge, 1984)

Lawrence, T. E., *The Seven Pillars of Wisdom*: *The Complete 1922 'Oxford' Text*, ed. J. and N. Wilson (Fordingbridge, Hampshire, 2004)

Leese, Arnold Spencer, *The One-Humped Camel in Health and Disease* (Stamford, 1927)

—, *Out of Step: Events in the Two Lives of an Anti-Jewish Camel-doctor* (Guildford, 1951)

Leonard, Arthur Glyn, *The Camel, Its Uses and Management* (London, 1894)

Lyall, Charles James, ed. and trans., *The Mufadaliyyat* (London, 1924)

—, 'Translations of Ancient Arabian Poetry, Chiefly Pre-Islamic' (London, 1930)

Macaulay, Rose, *The Towers of Trebizond* (London, 1956)

Peter McGrath, 'Sudan's Camel Healers', in *Dry: Life Without Water*, ed. Ehsan Masood and Daniel Schaffer (Cambridge, MA, 2006) pp. 16–25

Mahmoud, Neffi Bel-Hajj, *La Psychologie des animaux chez les Arabes* (Paris, 1977)

Marozzi, Justin, *South from Barbary: Along the Slave Routes of the Libyan Sahara* (London, 2001)

Martin, P. S., and H. E. Wright, Sr, *Pleistocene Extinctions: The Search for a Cause*, vol. VI of *Proceedings of the Seventh Congress of the International Association for Quaternary Research* (New Haven, CT, 1967)

McGrath, Peter, 'Sudan's Camel Healers', in *Dry: Life Without Water*, ed. Masood and Schaffer

Moorhouse, Geoffrey, *The Fearful Void* (London, 1974)

Musil, Alois, *The Manners and Customs of the Rwala Bedouins* (London, 1974)

Nelson, Cynthia, ed., *The Desert and Sown: Nomads in the Wider Society* (Berkeley, CA, 1873)

Norwich, John Julius, *Sahara* (London, 1968)

Pellat, Charles, *Encyclopedia of Islam*, 2nd edn, s.v. Ibil.

Prothero, Donald, *After the Dinosaurs: The Age of Mammals* (Bloomington and Indianapolis, 2006)

Ripinsky, Michael M., 'The Camel in Ancient Arabia', *Antiquity*, XLIX (1975), pp. 295–8

—, 'Camel Ancestry and Domestication in Egypt and the Sahara', *Archaeology*, XXXVI (1983), pp. 21–7

Robinson, A. E., 'The Camel in Antiquity', *Sudan Notes and Records*, XIX (1936), pp. 47–69

Romer, Alfred S., *The Vertebrate Story* (Chicago, 1959)

Roux, Jean-Paul, 'Le Chameau en Asie Centrale: son nom – son elevage – sa place dans la mythologie' *Central Asiatic Journal*, V (1959), pp. 35–76

Rowland, Beryl, *Animals with Human Faces: A Guide to Animal Symbolism* (London, 1974)

Ruskin, John, *The Works of John Ruskin*, ed. E. T. Cook and Alexander Wedderburn, 39 vols (London and New York, 1903–12)

Russell Pasha, Thomas, *Egyptian Service 1902–1946* (London, 1949)

Salzman, Philip Carl, ed., *When Nomads Settle: Processes of Sedenterization as Adaptation and Response* (New York, 1980).

Schmidt-Nielsen, Knut, *Desert Animals: Physiological Problems of Heat and Water* (Oxford, 1964)

Screech, Timon, *Sex and the Floating World: erotic images in Japan, 1700–1820* (London, 1999)

Serjeant, R. B., review of Bulliet's *The Camel and the Wheel*, in *Bulletin of the School of Oriental and African Studies*, XXXIX (1976), pp. 662–4

Sells, Michael, 'Mu'allaqa of Tarafa translated by Michael Sells', *Journal of Arabic Literature*, XVII (1986), pp. 21–34

Smuts, Malie M. S. and A. J. Bezeidenhout, *Anatomy of the Dromedary* (Oxford, 1987).

Stetkevych, Suzanne Pinckney, *Mute Immortals Speak: Pre-Islamic Poetry and the Poetics of Ritual* (Cornell, NY, 1973)

Stetkevych, Jaroslav, 'Name and Epithet: The Philology and Semiotics of Animal Nomenclature in Early Arabic Poetry', *Journal of Near Eastern Studies*, XLV (1986), pp. 89–124

Sweet, Louise E., 'Camel Raiding of the North Arabian Bedouin; A Mechanism of Ecological Adaptation', *American Anthropologist*, LXVI (1965), pp. 1132–50

Tate, Ro, *Desert Animals* (London, 1971)

Tench, Richard, *Forbidden Sands* (London, 1978)

Thesiger, Wilfred, *Arabian Sands* (London, 1959)

Van Gelder, Geert Jan, *Of Dishes and Discourse: Classical Arabic Literary Representations of Food* (Richmond, Surrey, 2000)

Von Dumreicher, André, *Trackers and Smugglers in Egypt* (London, 1951)

Von Hammer, Joseph Freiherr von, 'Das Kamel', *Kaiserliche Akadamie Wissenschaft*, vols VI and VII (1855–6)

Von Sivers, Peter, review of Bulliet's *The Camel and the Wheel* in *Journal of the American Oriental Society*, IX (1978), pp. 163–4.

War Office, *Camel Corps Training Provisional* (London, 1913)

Wellard, James, *Samarkand and Beyond: A History of Desert Caravans* (London, 1977)

Wilson, R. T., *The Camel* (London, 1984)

—, 'The Nutritional Requirements of Camel', *Options Méditerranéennes*, II (1989), pp. 171–9

Wolseley, Sir Garnet J., *The Soldier's Pocket-Book for Field Service*, 4th edn (London, 1882)

Wood, Frances, *The Silk Road: Two Thousand Years in the Heart of Asia* (London, 2002)

Yagil, Reuven, *The Desert Camel: Comparative Physiological Adaptation* (Basel, 1985)

Associations and Websites

There are an astonishing number of websites devoted to camels. Only a few of the most useful will be listed here:

AUSTRALIAN CAMEL
http://austcamel.com.au/inform.htm

THE A–Z OF CAMELS
www.arab.net/camels

CAMELGATE
('the gateway to knowledge of sustainable use of camels')
www.camelgate.com

THE CAMEL HUB OF THE WEB
www.camelphotos.com

CAMELS AND THE NABATAEANS
http://nabataea.net/camel.html

IMPARJA CAMEL CUP
www.camelcup.com.au

JOURNAL OF CAMEL PRACTICE AND RESEARCH,
www.camelsandcamelids.com

LIVIUS.ORG
(devoted to the camel in classical antiquity)
http://www.livius.org/caa-can/camel/camel.html

LOKHIT PASHU-PALAK SANSTHAN
(devoted to the welfare of Rajasthan's pastoralists)
www.lppps.org

NATIONAL RESEARCH CENTRE ON THE CAMEL, BIKANER, RAJASTHAN,
INDIA
www.icar.org.in/nrccm/index.php

WILD CAMEL PROTECTION FOUNDATION
(the main site for the protection of the wild Bactrian. Jane Goodall is
the patron of the Foundation and John Hare its founder and chair-
man.)
www.wildcamels.com/book2html

WILD CAMELS
www.wildcamels.com/news18htm

Acknowledgements

Grateful acknowledgements, first to Neil Hornick (my 'tireless legman'), then Helen Irwin (who took many of the photographs), Ted Irwin, John Hare of the Wild Camel Foundation and Oliver Duprey, formerly of the London Zoo. Also Juri Gabriel, John Baxter, Richard Bulliet, William Clarence-Smith, Kevin Jackson, Alastair Hamilton, Brian Dunn, Barnaby Rogerson, Ann Sylph, the Arcadian Library, the staffs of the library of the London Zoological Society, the London Library and the library of School of Oriental and African Studies.

Also the people at Reaktion who made the production of this book so pleasant, especially Michael Leaman, Harry Gilonis and Martha Jay.

Photo Acknowledgements

The author and publishers wish to express their thanks to the following sources of illustrative material and/or permission to reproduce it. Locations, etc., of some items not placed in the captions for reasons of space are also given below.

Photo Alinari/Colorpoint/Rex Features: p. 6; from Manó Andrásy, *Utazás Kelet Indiákon, Ceylon, Java, Khina, Bengál* (Budapest, 1853): p. 83; reproduced courtesy of David B. Appleton/Appleton Studios: p. 111; photo courtesy of the The Arcadian Library, London: p. 117; Art Institute of Chicago: p. 144; photos courtesy of the author: pp. 49, 168, 171, 176, 179, 181, 189; collection of the author: pp. 27, 71; Bibliothèque Nationale de France, Paris: p. 80; Bodleian Library, University of Oxford: p. 109; British Library, London: p. 110; British Museum, London (photos © Trustees of the British Museum): pp. 72, 89, 93, 96, 141, 175 (foot), 184; from le Comte de Buffon, *Histoire Naturelle . . .* vol. 29 (Paris, 1799–1800): p. 8; Capella degli Scrovegni, Padua: p. 113; photo click/ morgueFile.com: p. 15; photo Daily Mail/Rex Features: p. 174; The David Collection, Copenhagen: p. 91; photos courtesy of Oliver Duprey: pp. 13, 133; photo courtesy of Graber AG, Switzerland (www.kagra.ch): p. 180; from John Hare, *The Lost Camels of Tartary* (London: Little, Brown, 1998), by permission of the author (John Hare): p. 191; El Jem Museum, El Jem, Tunisia: p. 145; photo Philip Jones: p. 54; Nasser D. Khalili Collection of Islamic Art, © the Nour Foundation, courtesy the Khalili Family Trust: p. 75; from Rudyard Kipling, *Just So Stories* (London, 1902): p. 129; photos Pernille Klemp: pp. 77, 91; Kunstindustrimuseet,

Copenhagen: p. 77; courtesy Jean-Marie Le Tensorer: pp. 44, 45; Library of Congress, Washington, DC (Prints and Photographs Division): pp. 22, 51, 53 (all G. Eric and Edith Matson Photograph Collection), 102 (Chadbourne Collection of Japanese Prints), 167 (Frank and Frances Carpenter Collection); © Michael Long/Natural History Museum, London Picture Library: pp. 38, 40, 42; Metropolitan Museum of Art, New York: pp. 84, 108; Musée d'Orsay, Paris: p. 86; Musée du Louvre, Paris: pp. 76, 98 (foot), 112, 115 (top); from Alois Musil, *The Manners and Customs of the Rwala Bedouins...* (New York, 1928): pp. 50, 81, 175 (top); National Palace Museum, Taipei, Taiwan: p. 106; photo K. Nomachi/ Rex Features: p. 66; private collections: pp. 99, 103, 104, 115 (foot), 164; Clifford R. Prothero: p. 39; from Donald R. Prothero, *After the Dinosaurs* (Bloomington, IN: Indiana University Press, 2006), reproduced by kind permission of Indiana University Press: p. 39; from Eadweard Muybridge, *Animal Locomotion* (Philadelphia, 1887): p. 24; photo Lori Parsons: p. 25; photo © ralphp/2010 iStock International Inc.: p. 59; from *reCollections (Journal of the National Museums of Australia)* vol. 2, no. 2 (September 2007): p. 54; from *Report of the Secretary of State for War, communicating, in compliance with a resolution of the Senate of February 2, 1857, information respecting The Purchase of Camels for the purposes of military transportation* (Washington, DC, 1857): pp. 52, 57, 61, 149; Peter Schmid, University of Zurich: pp. 44, 45; from Edward Topsell, *The Historie of Foore-Footed Beastes* (London, 1607): p. 158; Victoria & Albert Museum, London London (photos V&A Images): pp. 18, 98 (top), 105; from Arnold von Harff, *The pilgrimage of Arnold von Harff, knight, from Cologne through Italy, Syria, Egypt, Arabia, Ethiopia, Nubia, Palestine, Turkey, France, and Spain, which he accomplished in the years 1496 to 1499*: p. 156; from Elijah Walton, *The Camel* (London, 1865): p. 117; Werner Forman Archive/al-In Museum, Dubai: p. 148; Werner Forman archive/Biblioteca Nacional, Madrid: p. 152; from R. T. Wilson, *The Camel*, © Pearson Education Ltd, 1984: pp. 20, 21, 28; Yale Centre for British Art, New Haven, CT (Paul Mellon Collection): p. 119; photos Zoological Society, London: p. 29, 55.

Index

227